TEMPORAL LOGICS
and
their applications

TEMPORAL LOGICS
and
their applications

Edited by

Antony Galton

Department of Computer Science
University of Exeter

1987

ACADEMIC PRESS
Harcourt Brace Jovanovich, Publishers
London · San Diego · New York · Berkeley
Boston · Sydney · Tokyo · Toronto

ACADEMIC PRESS LIMITED
24-28 Oval Road, London NW1 7DX

United States Edition published by
ACADEMIC PRESS, INC.
San Diego, CA 92101

British Library Cataloguing in Publication Data

Temporal logics and their applications.
1. Electronic digital computers—
Programming   2. Logic, Symbolic and
mathematical
I. Galton, A.P.
005.13'1   QA76.6

ISBN 0-12-274060-2
LCCN 87-71534

Typeset by Eta Services (Typesetters) Ltd, Beccles, Suffolk

Printed in Great Britain by
St Edmundsbury Press, Bury St Edmunds, Suffolk

CONTRIBUTORS

H. BARRINGER
Department of Computer Science, The University, Manchester M13 9PL, UK

D. M. GABBAY
Department of Computing, Imperial College of Science and Technology, University of London, 180 Queen's Gate, London SW7 2BZ, UK

A. P. GALTON
Department of Computer Science, University of Exeter, Prince of Wales Road, Exeter EX4 4PT, UK

R. W. S. HALE
University of Cambridge Computing Laboratory, Corn Exchange Street, Cambridge CB2 3QG, UK

F. SADRI
Department of Computing, Imperial College of Science and Technology, University of London, 180 Queen's Gate, London SW7 2BZ, UK

FOREWORD

Computer science is the study of algorithms, programming languages for the expression of algorithms, and computer systems for the implementation of programming languages. An important activity in computer science is the invention, analysis and application of formally defined calculi, called *logics*, which are designed to specify, reason about, and represent these algorithms, programs and systems.

The articles in this book concern logics for reasoning about time, and illustrate their application in all three areas of computer science. For example, research on the logic of events is brought together with research on the specification and verification of concurrent computations, and research on programming constructs for information systems. I believe this volume is the first to present new technical work on temporal logic from philosophy, software engineering and artificial intelligence; the inclusion of material from artificial intelligence is particularly useful. A coherent account of these fields is given in Dr Galton's excellent Introduction.

Temporal logic is one of several theories (not all of them mathematical theories) that have arisen in philosophy and are finding application in computer science. I think the case of temporal logic exemplifies rather well how progress in an academic field can depend on the importation and adaption of ideas from an apparently unrelated field. We are reminded that intellectual life is not to be regimented; also that speculative research is a precious possession of our universities. Let it not go unnoticed by the readers of this book that for the past few years, while computer science has received every encouragement, philosophy has been discouraged and dismantled in many universities. I think that, in less than a decade, this disregard for philosophy will be seen as unforgiveable stupidity.

The book is based on a conference organised by Dr Galton on behalf of the *Centre for Theoretical Computer Science* at the University of Leeds. The aim of the Centre, like that of the meeting and the book, is to help create an interdisciplinary *milieu* in which to experiment with research relevant to the foundations of computer science. I hope the reader will enjoy the different contributions and will be inspired by the variety of problems they raise.

J. V. Tucker
Director of the Centre for Theoretical Computer Science
University of Leeds

PREFACE

This book has arisen from a conference on Temporal Logic and its Applications held at the University of Leeds in January 1986 under the auspices of the then newly created Centre for Theoretical Computer Science. Some sixty delegates attended the conference, drawn mainly from computer science, philosophy, and mathematics. Of the papers in this book, those by Barringer, Galton, and Hale are directly based on presentations given at the conference, the other papers being also closely related to material presented there.

As the reader of this book will discover, temporal logic is a field which, having originated within philosophy, has now proved to be of relevance to several distinct areas in computer science. This is, I believe, the first publication in which all of these aspects of temporal logic are treated together. It is to be hoped that the book will provide a stimulus to further inter-disciplinary collaboration, not only as regards temporal logic itself but also in connection with other logical and philosophical issues which lie at the interface between computing and philosophy.

Each of the chapters in the book is entirely self-contained, and can be read independently of all the others. It is recommended, however, that the reader who is unfamiliar with temporal logic in any form should first read the introductory chapter.

Antony Galton

CONTENTS

1 TEMPORAL LOGIC AND COMPUTER SCIENCE: AN OVERVIEW
Antony Galton

2 THE USE OF TEMPORAL LOGIC IN THE COMPOSITIONAL SPECIFICATION OF CONCURRENT SYSTEMS
Howard Barringer

3 TEMPORAL LOGIC PROGRAMMING
Roger Hale

4   THREE   RECENT   APPROACHES   TO   TEMPORAL   ⊛
    REASONING
    Fariba Sadri

5   THE LOGIC OF OCCURRENCE
Antony Galton

6   MODAL AND TEMPORAL LOGIC PROGRAMMING
Dov Gabbay

1 TEMPORAL LOGIC AND COMPUTER SCIENCE: AN OVERVIEW

Antony Galton

*Department of Computer Science,
University of Exeter*

INTRODUCTION

In recent years it has increasingly been argued that philosophers can no longer afford to ignore the fundamental innovations in concept and method introduced by practitioners of various disciplines in Computer Science, notably in the study of Artificial Intelligence. Work in many of the traditional problems of philosophy, particularly those associated with language and mind, ought, according to this line of argument, to receive fresh impetus from the insights of those who have sought to tackle related problems from a practical, computational viewpoint (see, for example, Sloman, 1978).

Whatever one may think of the specific claims of Aaron Sloman and others who have argued along these lines, one cannot seriously deny that philosophy always has drawn inspiration from close association with other intellectual disciplines, and there is no reason to suppose that computer science should prove an exception in this regard; there are already signs of growing interest in all matters computational within the philosophical community, and there is every expectation that this trend will continue into at least the immediate future. Conversely, it is just as clear that computer science has a good deal to learn from philosophy, and indeed computer scientists are showing ever greater awareness of and interest in those aspects of philosophy with the greatest relevance to their own discipline.

Temporal Logic affords a particularly good illustration of the sort of close association between two disciplines I have alluded to. It arose, in the form of Tense Logic, purely within the domain of philosophy, and was pursued for many years by philosophers and logicians without any regard to its possible applications outside their subject. Yet over the last decade or so it has become apparent that temporal logic has valuable potential as a formal tool for use in

Temporal Logics and their Applications

ISBN: 0-12-274060-2

both artificial intelligence and software engineering. This has come about, I think, because computer science as a whole is at once highly formal and deeply rooted in the practicalities of everyday life, so that a formalism designed to handle so pervasive a feature of everyday life as time has a natural role to play in it.

This introductory survey falls into three main sections. In the first, I examine the philosophical background in which temporal logic was first developed; next, I consider the applications of ideas from temporal logic to the fields of Artificial Intelligence and Cognitive Science, in which temporal logic is used to enable computer programs to reason about the world; and lastly, I examine the further development of temporal logic in software engineering as a tool to enable the world to reason about computer programs.

1 THE PHILOSOPHICAL BACKGROUND

1.1 The First-order Approach to Time

In mathematics, time has traditionally been represented as just another variable, indeed time is *the* independent variable *par excellence*, the dependence of a variable on time being the paradigmatic example of the functional dependence of one quantity on another. This is because historically one of the most important activities of mathematicians has been that of giving a formal quantitative account of physical processes, and such processes by their very nature always involve time. In physics, times are generally represented by real numbers, but in other applications integers or yet other mathematical objects may be more appropriate. Any such model of time involves the use of individual terms to designate times, and the logical apparatus employed for reasoning with such terms is invariably the first-order predicate calculus. Thus far, then, there is no need to speak of *temporal* logic; if times are designated by terms in a first-order theory, then this is just ordinary logic applied to a subject matter which happens to include time. And this is adequate for physics and the other mathematical sciences.

Since Frege, however, philosophers have been interested in logic particularly in its relation to language. Here, unlike in applied mathematics, time does not appear in the guise of 'just another variable'. We do, to be sure, have terms for designating particular times—clock times, calendar dates, and so on—but this way of encoding temporal information looks from a linguistic perspective far less fundamental than the stock of temporal connectives ('when', 'while', 'before', 'after', 'since', and 'until') and adverbs ('now', 'then', 'always', 'sometimes', 'never', 'soon', etc.) which form some of the most basic vocabulary of every language. And of course temporal notions are tightly woven into the *grammatical* fabric of most languages through the tenses of the verb.

Despite this, the first modern attempts to subject the temporal element of language to a logical treatment inherited, apparently uncritically, the mathematical paradigm of first-order terms denoting individual times; and all temporal features of language were regimented into the forms available within this paradigm.

Thus we find Bertrand Russell defining *change* as 'the difference, in respect of truth or falsehood, between a proposition concerning an entity and a time T and a proposition concerning the same entity and another time T', provided that the two propositions differ only by the fact that T occurs in the one where T' occurs in the other.' (Russell, 1903, §442). It is important to note here that Russell speaks of the truth or falsity of a proposition concerning an entity *and* a time—in other words of the eternal truth-value of a proposition of the form $f(x, t)$—rather than of the truth or falsity of a proposition concerning an entity *at* a time—which would be the temporary truth-value, at t, of a proposition of the form $f(x)$.

Quine, who thoroughly endorses this Russellian style of analysis, offers such transcriptions as

$$\neg(\exists x)(x \text{ is a time} \wedge \text{Jones is ill at } x)$$

for 'Jones is never ill' (Quine, 1965, p. 91), and goes a step further than Russell, replacing the latter's idea that properties should be construed as relations between entities and times by the idea that they should be construed still as properties, but properties of temporal stages of entities, substituting something like '$f(x$–at–$t)$' for '$f(x, t)$'. Here x–at–t is merely a 'slice', at right angles to the time axis, of the extended four-dimensional object x.

1.2 The Modal Approach

The methods of analysis proposed by Russell and Quine effectively remove time from the domain of logic, by treating temporal elements of a sentence on a par with other non-temporal elements. The first suggestion that a logical calculus might be constructed specifically in order to handle inferences involving time, thereby according temporal elements logical status, appears to have been made by Findlay in 1941. Findlay's suggestion arose in connection with a critical assault on McTaggart's notorious proof of the unreality of time; but it was not taken up until the early 1950s, when Prior constructed the first *tense logic*.

Prior had been impressed by a remark in a book review by Geach to the effect that the '*p* at *t*' style of analysis was quite alien to a discussion of ancient and mediaeval philosophy, which knew nothing of such analyses. Now expressions like '*p* at *t*' *had* been used by Benson Mates in his discussion of some theories of the Stoic logician Diodorus Chronus, and Prior hoped to be able to recast Mates's discussion in terms not involving the '*p* at *t*' style of analysis which Geach had stigmatized as anachronistic.

Diodorus apparently sought to reduce *modal* notions to *temporal* ones, defining the *possible* as 'what is or will be', the *necessary* as 'what is and always will be', and the *impossible* as 'what is not and never will be'. Prior was interested to discover which of the existing formal systems of modal logic was implicit in Diodorus's definitions, and realised that, in his own words, 'to get a logic of the possible from its definition in terms of the future, one must also have a *logic* of futurity. The construction of a calculus of tenses could not wait much longer.' (Prior, 1967, p. 17).

The seminal paper in which Tense Logic first saw the light of day, Prior (1955), already contained the now-familiar notations Fp (for 'It will be the case that *p*'), Gp (for 'It will always be the case that *p*'), Pp (for 'It has been the case that *p*'), and Hp (for 'It has always been the case that *p*'), as well as the identifications of Gp and Hp with $\neg F \neg p$ and $\neg P \neg p$ respectively.

In this initial paper, Prior showed that, given a number of intuitively plausible postulates concerning the logic of futurity, the Diodorean modal system obtained by defining

$$\Diamond p =_{\text{def}} p \vee Fp$$

after the fashion of Diodorus is at least as strong as the Lewis system S4, but weaker than S5. To get a system as strong as S5, it is necessary instead to extend the Diodorean account of possibility by including the past as well as the future, i.e.

$$\Diamond p =_{\text{def}} Pp \vee p \vee Fp.$$

This is not the place to explore the detailed ramifications of this initial work on the relationship between tense and modality; an excellent account can be found in Prior (1967), especially Chapter 2. Suffice it to say that because of the circumstances attending its birth, much of the early work in Tense Logic was devoted to articulating this relationship in detail. In fact, Tense Logic is generally regarded as a *kind* of modal logic, differing from ordinary modal logics in possessing two sets of operators, one for the past and one for the future, where ordinary modal logic has only one. And so the Prior style of analysis of temporal discourse has come to be known as the *modal* approach to time, as distinct from the *first-order* approach typified by Russell and Quine.

1.3   Tense Logic and the Topology of Time

It quickly became apparent to Prior and other workers in tense logic that the new formalism was potentially a valuable tool for expressing, and exploring the consequences of, a range of theses about the structure of time, theses about what in a somewhat loose way of speaking has come to be called the *topology* of time. Examples of questions which have been considered are:

(a) Is time *discrete* or *continuous?*
This question can be taken at the ontological level as 'Are there really time atoms?' or as a question about suitable representations within a given domain—for example, if one is dealing with facts which only vary on a time-scale consisting of whole numbers of days, then one can take a day as an indivisible unit and use a discrete time model. Again, suppose we have *n* property-variables, each of which is present or absent at any one time, and that we define an *epoch* to be a maximal period over which none of the variables changes value. Then the succession of epochs constitutes a discrete time structure imposed on what may very well be an intrinsically continuous underlying temporal framework.
   Note that if time is treated as discrete, then we can introduce a new operator $\bigcirc$, read 'at the next moment'. This operator is widely used in the computer science applications to be described below.

(b) Is time *bounded* or *unbounded?*
Note that this is not quite the same as the question whether time is finite or infinite in extent. The latter question only has meaning if a metric has been defined on the time line. But even without a metric, there are still the possibilities that either every time is succeeded by a later time (unbounded in the future) or there is a last moment in time (bounded in the future), and likewise in the direction of the past.

(c) Is time *linear, parallel,* or *branching?*
Here we are dealing with the gross topological properties of time. The idea of branching time has been put forward as a way of handling uncertainty about the future, the idea being that from each moment forward there exist many possible alternative futures. Unfortunately, this picture does not succeed very well in capturing the idea that of all the possible alternative futures there is one privileged 'actual' future. In the branching time systems, one can easily express the propositions that it *may* be going to be the case that *p* and that it is *bound* to be the case that *p*, but it is not so easy to express the proposition that it *will* be the case that *p*: so a branching time model is only likely to

appeal to someone who does not think this third proposition has any meaning distinct from the other two.

Against this, there seems to be something very compelling about the idea of the actual future, distinct from all other possible futures. To accord actuality to all the branches does not seem to be very meaningful unless there is some way of communicating between the branches so that they can interact—this is the stuff of science fiction, as for example in some of the short stories of John Wyndham—and not something that many philosophers, at least, would be prepared to take seriously.

As for *parallel* times: we could perhaps regard this as a model of the different subjective time-scales of different people, but again, in order to set up a formal calculus here it would seem necessary to establish correlations between the different times, and the robustness of such correlations is a measure of how well-founded is the idea of an *objective* time underlying all the different subjective time-lines.

Circular time is an interesting possibility, too, for in this, past, present and future coalesce. It is hard to take circular time seriously as an account of physical time, but the associated logic could be useful in reasoning about repetitive processes, for example the effectively endless repetition of cycles in the traffic signals at a road junction.

All these topological issues are discussed at length by Newton-Smith (1980).

1.4   The Two Approaches—Rivals or Allies?

The modal and first-order approaches may be viewed either as rivals or as allies.

1.4.1   The two approaches as rivals

There have been extensive philosophical debates as to which approach is 'correct'. Massey went so far as to speak of a Kuhnian paradigm clash (Massey, 1969). Massey invented the now popular terminology of 'tensers' and 'detensers' to describe the adherents of the modal and first-order approaches respectively. What are the arguments?

(a) *Metaphysical arguments about existence and time.* On the tenser view there is a crucial distinction between such pairs of sentences as

$$F(\exists x)f(x)$$

and

$$(\exists x)Ff(x)$$

(i.e. 'It will be the case that there is something which is f' and 'There is something of which it will be the case that it is f'). The former sentence might be true if something that *does not yet exist* is later on going to exist and be f, whereas the latter can only be true if there *already exists* something that is later going to be f.

For the detenser, these formulae come out as something like

$$(\exists t)[\text{later}(t, \text{now}) \wedge (\exists x)f(x, t)]$$

and

$$(\exists x)(\exists t)[\text{later}(t, \text{now}) \wedge f(x, t)]$$

and these are easily seen to be equivalent in first-order logic; so for the detenser the alleged distinction cannot be upheld. This means that tensers (or at least those who accept the distinction) and detensers must take a radically different view of the relationship between existence and time. For the detenser, anything which exists exists timelessly; its temporal qualifications are a result of its standing timelessly in relation to certain times. In this view of the world, time is treated as a dimension on a par with the three dimensions of space; it is generally felt that the theory of Relativity lends support to this way of regarding time, although the latter is in no way dependent on the former. For the tenser, this 'four-dimensionalism' is, if not simply incoherent, then at least radically deficient in that it leaves totally out of account that *transience* which, according to the tenser, is an essential feature of time and temporal phenomena.

(b) *Conceptual arguments.* A different line that is taken by tensers is that if we want our temporal logic to mirror *conceptual* priorities—which from a philosophical point of view is surely desirable—then we must treat tenses and all the other adverbial paraphernalia of temporal reference as more basic than dates and times. An eloquent exponent of this point of view has been Geach, who argued that it is perverse to try to analyse 'grass roots temporal discourse' (i.e. tenses and simple temporal relations like 'before' and 'after') in terms of the 'vastly more complex notions' implicit in the first-order approach (Geach, 1965).

Of course, even if one grants that Geach's point is a compelling one from the point of view of ultimate philosophical analysis of the structure of our temporal concepts, the detenser may still with some propriety claim that that point of view is not the only one in which there is need to formalize temporal language, and that from the point of view of physics, say, or computer databases, it would be hopelessly cumbersome to analyse all times and dates into Geach's 'grass-roots' style of discourse, even if it could be agreed just what is the correct way of doing this.

(c) *Logical arguments.* There is a persistent undercurrent in much modern logic to the effect that first-order logic is in some way definitive, that nothing is truly intelligible unless cast in first-order form. This view has been championed by Quine and by Davidson, amongst others (in Susan Haack's words, for Quine 'extensionality is the touchstone of intelligibility'—the point being that first-order logic is extensional, whereas modal logics are not). First-order treatments have the advantage that first-order logic is known to be complete with respect to its standard model theory (as against, e.g. second-order logic—although necessity compels us to continue doing arithmetic!), and there exist decision procedures and proof procedures for substantial and important subsets of it (even though the theory as a whole provably lacks either). Moreover, for computer science, first-order treatments lend themselves readily to implementation as logic programs, especially if they are expressed entirely in terms of Horn Clauses, in which case the implementation in Prolog is practically automatic.

(d) *Expressive power.* Not everything that can be expressed in the first-order approach is expressible in the modal approach, unless additional operators are introduced. On the other hand, the modal approach has the advantage that its expressions are both simpler and closer in spirit to natural language. This issue is treated in greater detail below (pp. 39–41).

1.4.2   The two approaches as allies

The alliance between the two approaches[1] comes about because the first-order approach can be made to serve as a *model theory* for the modal approach, i.e. in order to give a formal account of the intended meanings of the operators in the latter approach, one typically makes use of formalisms drawn from first-order logic.

This state of affairs is almost inevitable. If asked, what does it mean to say 'It will be the case that *p*', it is very hard to avoid coming up with something like 'There is some future time at which *p* is true'. A tenser might say that this is a mere *façon de parler*, but it is nonetheless impressive that we are able in everyday language to handle quantification over times in a way quite analogous to the way in which we handle quantification over less contentious entities. and without, apparently, lapsing into gross conceptual confusion. Even Prior, the arch-tenser, has to resort to this sort of talk to explain precisely the meaning of formulae in tense logic.

The style of model theory standardly used for Tense Logic is a straightforward adaptation of the well-known Kripke semantics for modal

---

1. The discussion in this section owes much to van Benthem (1983).

logics (Kripke, 1963), originally due to Kanger (1957). It is noteworthy that this particular semantical paradigm seems a good deal more natural when it is applied to tense logics than when it is applied to modal logics, since of the entities invoked in each case, *times* are a part of our everyday gamut of concepts in a way that *possible worlds* (which play the corresponding role in the modal semantics) just aren't.

It should be noted that Prior himself had little to do with the model theory of tense logic, largely preferring to confine his technical discussion to the proof theory. Accounts of the model theory can be found in Rescher and Urquhart (1971) and in McArthur (1976). Rescher and Urquhart also give proof systems for both linear and branching tense logic based on the method of semantic tableaux. Alongside Prior (1967), Rescher and Urquhart's book has assumed the status of a classic in the literature on temporal logic; it remains to be seen whether the recent scholarly and comprehensive book by van Benthem (1985) will achieve a similar distinction.

The formal semantics for Tense Logic is based on the following 'translations':

$$Pp \text{ is true at } t \Leftrightarrow (\exists t')[\text{later}(t, t') \wedge p(t')]$$

$$Fp \text{ is true at } t \Leftrightarrow (\exists t')[\text{later}(t', t) \wedge p(t')]$$

$$Hp \text{ is true at } t \Leftrightarrow (\forall t')[\text{later}(t, t') \rightarrow p(t')]$$

$$Gp \text{ is true at } t \Leftrightarrow (\forall t')[\text{later}(t', t) \rightarrow p(t')]$$

Using these, we can straightforwardly convert any tense-logical formula into first-order form, e.g. the formula

$$p \rightarrow GPp$$

becomes

$$p(t) \rightarrow (\forall t')[\text{later}(t', t) \rightarrow (\exists t'')[\text{later}(t', t'') \wedge p(t'')]].$$

Since the latter is a theorem of first-order logic, the tensed version had better be a theorem of tense logic too, and indeed it, or something equivalent, is included as an axiom in all systems of tense logic.

Other formulae which translate into first-order theorems are $p \rightarrow HFp$, $G(p \rightarrow q) \rightarrow (Gp \rightarrow Gq)$, and $H(p \rightarrow q) \rightarrow (Hp \rightarrow Hq)$. This list is complete in the following sense: if we take them as axiom schemas for a tense-logical system, together with the tautologies of the propositional calculus, and if we posit as rules of inference

MP: from $\vdash \alpha$ and $\vdash \alpha \rightarrow \beta$ infer $\vdash \beta$

RG: from $\vdash \alpha$ infer $\vdash G\alpha$

RH: from $\vdash \alpha$ infer $\vdash H\alpha$

then the theorems of the resulting system are precisely those tense-logical formulae which translate into theorems of first-order logic. The system of tense logic here outlined is known as *Minimal Tense Logic*, and forms the basis for most other systems that have been studied.

Any tense-logical formulae that are not theorems of Minimal Tense Logic will translate into first-order formulae that are no longer theorems of first-order logic, for example the tense-logical formula

$$FFp \rightarrow Fp \tag{1}$$

which says that whatever will be future is already future, translates into the first-order formula

$$(\exists t')[\text{later}(t, t') \wedge (\exists t'')[\text{later}(t', t'') \wedge p(t'')] \rightarrow$$
$$(\exists t'')[\text{later}(t, t'') \wedge p(t'')] \tag{2}$$

which is not a first-order theorem.

If (1) is to be added as an *axiom schema*, thus extending Minimal Tense Logic, the consequence in the first-order translation is that (2) must be stipulated to hold for all times t and for all predicates p (since if (1) is to be treated as an axiom schema, p must be regarded as a schematic letter doing duty for any formula that might be substituted for it), so that in effect the translation of (1) *when regarded as an axiom schema* is now the *second-order* formula

$$(\forall p)(\forall t)[(\exists t')[\text{later}(t, t') \wedge (\exists t'')[\text{later}(t', t'') \wedge p(t'')]] \rightarrow$$
$$(\exists t'')[\text{later}(t, t'') \wedge p(t'')]] \tag{3}$$

which again fails to be a theorem of second-order logic.

In general, then, the addition of further axioms to Minimal Tense Logic amounts to the assertion of substantial second-order *theses* about the earlier-later relation. In the example cited, it turns out that this second-order thesis is in fact equivalent to a first-order one, namely

$$(\forall t)(\forall t')(\forall t'')[(\text{later}(t, t') \wedge \text{later}(t', t'')) \rightarrow \text{later}(t, t'')] \tag{4}$$

which asserts the transitivity of the relation 'later'. For the most part, it is first-order statements like (4) rather than second-order ones like (3) that are of interest to us, for they correspond to intuitively 'salient' properties that a temporal ordering may possess, such as transitivity, density, boundedness, and so on. It is therefore of interest to know which tense-logical formulae, when taken as axiom schemas, yield first-order properties.

Not all do: a simple example which does not is McKinsey's formula $GFp \rightarrow FGp$. Van Benthem (1983, p. 158) has given an exact characterization of those tense-logical formulae which, when taken as axioms, give rise to first-order formulae. This tells us which formulae are first-order definable; but there still does not exist a general effective procedure for obtaining these first-order equivalents when they exist.

The converse task, that of finding a tense-logical formula to correspond to a given first-order formula, is also of interest, because it bears rather strongly on the issue of the relative expressive powers of the two methods of representation. In fact it is not hard to find first-order properties of temporal orderings which are not tense-logically definable, a well-known example being the formula

$$(\forall t)\neg \operatorname{later}(t, t)$$

which expresses the *irreflexivity* of the temporal ordering. This means, in effect, that no tense-logical system can constrain its models to be irreflexive.

Once again, it can be shown that there is a precisely defined class of transformations such that a first-order sentence making use only of the predicates 'later' and '=' is tense-logically definable if and only if it is preserved under that class (see van Benthem 1983, p. 161).

1.5 Some Related Work on the Logic of Temporal Discourse

The fact that there are first-order formulae in 'later' and '=' which are not tense-logically expressible means that the language of tense logic is *expressively incomplete* with respect to the first-order temporal logic. This naturally prompts the question whether the tense-language can be extended so as to become expressively complete in this sense. An affirmative answer is supplied by the work of Kamp (1968), who developed a temporal logic using the *binary* operators S and U, read 'since' and 'until', whose translations into first-order language are

$$Spq \text{ is true at } t \Leftrightarrow (\exists t')[\operatorname{later}(t, t') \wedge p(t') \wedge$$
$$(\forall t'')[(\operatorname{later}(t'', t') \wedge \operatorname{later}(t, t'')) \rightarrow q(t'')]]$$

$$Upq \text{ is true at } t \Leftrightarrow (\exists t')[\operatorname{later}(t', t) \wedge p(t') \wedge$$
$$(\forall t'')[(\operatorname{later}(t', t'') \wedge \operatorname{later}(t'', t)) \rightarrow q(t'')]].$$

Informally, Spq is true now so long as p has been true at some past time and q has been true ever since then; while Upq is true now so long as p will be true at some future time, and q will be true up to then. Kamp showed that, assuming continuity and strict linearity of the temporal order, *any* first-order temporal formula can be expressed as a formula of the S, U-calculus.

Note that it would be misleading simply to equate Kamp's operators 'S' and 'U' with the English conjunctions 'since' and 'until'. There are at least two major reservations one might have about such an equation. Consider an English sentence such as 'I will be unhappy until you come', and the putative rendering into Kamp's notation as Upq, where q is 'I am unhappy' and p is 'You come'. The first problem is that whereas Upq implies Fp, the English sentence does not imply 'You will come'; the second is that the English sentence suggests that once you *have* come, I shall no longer be unhappy, whereas this is obviously not an implication of Upq. Nonetheless, it does seem that S and U come as near to capturing the essence of 'since' and 'until' as any relatively simple logically defined pair of connectives could hope to do.

Other temporal connectives have also been studied by philosophers, with varying degrees of formality. Two classic treatments are Anscombe's largely informal study of 'before' and 'after', and von Wright's formal T-calculi, embodying a connective T, to be read as 'and then' if time is treated as continuous, and as 'and next' if time is discrete. Both these studies are of interest because they link up with other important areas in the study of temporal logic, Anscombe's with the logic of *aspect*, and von Wright's with the logic of *action*; and both these areas reach beyond the domain of tense logic proper because of the prominence assumed in them by the notion of an *event*.

1.5.1 Aspect

Anscombe (1964) noted that, contrary to what one might imagine, 'before' and 'after' are not strict converses of each other. For example, it is true that

$$Haydn\ was\ alive\ before\ Mozart\ was\ alive \tag{5}$$

and that

$$Haydn\ was\ alive\ after\ Mozart\ died, \tag{6}$$

but it is not true either that

$$Mozart\ was\ alive\ after\ Haydn\ was\ alive \tag{7}$$

or that

$$Mozart\ died\ before\ Haydn\ was\ alive; \tag{8}$$

whereas if 'before' and 'after' were strict converses, one would expect (5) to be equivalent to (7), and (6) to (8).

On the other hand, the sentences

$$Haydn\ was\ born\ before\ Mozart\ was\ born$$

and

$$Mozart\ was\ born\ after\ Haydn\ was\ born$$

do appear to be pretty nearly equivalent, suggesting that here 'before' and 'after' *are* converses. Anscombe noted that 'before' and 'after' are only true converses when the clauses they link report the occurrence of instantaneous events, and it is this observation more than any that links her work with the study of aspect.

Primarily, the term *aspect* refers to those elements of grammatical structure which draw attention to how a state of affairs or event is presented in a sentence, so that for example the *perfective* aspect (represented in English by the simple past tense as in 'I wrote a letter') presents an event as a completed whole, whereas the *imperfective* aspect ('I was writing a letter') presents it rather as something in progress, with no implication as to whether or not it eventually comes to completion. This simple description belies the complexity of what in fact has proved to be an enormously difficult and challenging area, which has increasingly during recent years attracted the attention of linguists, philosophers, and logicians.

For our present purposes, the most important idea is that the existence of aspectual distinctions gives rise to a taxonomy of verb-types according to their so-called *inherent aspect* or *aspectual character* (also known by the German term *Aktionsart*). Some verbs, or verb-phrases, refer to states (as in 'The cake is in the oven'), some to processes or activities (as in 'Mary is baking'), some to events which take time (as in 'Mary baked a cake'), and some to instantaneous events (as in 'The light went out'). This four-fold division was given by Vendler (1967), who labelled the four types 'states', 'activities', 'accomplishments', and 'achievements' respectively. Vendler was not the first to attempt a classification along these lines, and indeed, the beginnings of such a classification are already to be found in Aristotle (cf. Ackrill, 1965); but Vendler's classification seems to have been the most influential, and the variant schemes that have been worked out from time to time since then have seldom departed far from Vendler's own.

Much of the work on aspectual character has been, like Vendler's, discursive in nature, with little attempt to integrate the results into a formal system of temporal logic. Most of those who have attempted to do this have ended up rejecting the standard semantics for tense logic, which is based on instants, in favour of a semantics based on intervals. Interval semantics has been taken up by Cresswell (1977), Dowty (1979), Humberstone (1979), and Richards (1982).

The idea motivating the shift from instants to intervals appears to be that since some events typically take time, a sentence reporting an event ought to be assigned a truth-value, not relative to an instant (which is as it were too small to accommodate the whole event), but relative to an interval—the sentence being true on an interval just so long as the event it refers to takes up exactly that interval and no more.

I argued in Galton (1984) that the 'marriage of propositions with intervals is at best an unhappy one' (p. 21); it seemed to me that interval semantics arose from a confusion between something's being true *at* a time, and its being true *of* a time. It might well be true *of* a particular hour-long stretch, say, that I spent it writing a letter, but that does not mean that we should adopt a semantical scheme whereby the sentence 'I write a letter' is assigned the value 'true' relative to that hour-long stretch, for no such assignments can shed any light on what it is for a sentence about my letter-writing to be true *at the time that it is uttered*. Essentially the same point has been well argued, independently of my own work, by Pavel Tichý (in Tichý, 1985). For further discussion on this, see below (p. 171).

As an account of natural language semantics, then, interval semantics seems to me to be deficient. Despite this, there is, as we shall see, a use for an interval-based semantics of temporal logic within computer science (see below).

1.5.2 Action

Von Wright's *T*-calculi (von Wright, 1965, 1966) arose from an attempt to supply a formal basis for philosophical reasonings about *action*, this being what is required for a rational understanding of the notions of free will and responsibility and the ethical issues with which they are entangled. Much has been written, and continues to be written, in the attempt to elaborate a philosophically sound theory of human action, the key problem being perhaps one of definition: what is it about one set of bodily movements that makes it an action, while another apparently identical set is not?

Von Wright's work in this area focussed attention on the relationship between action and *change*. Some actions consist in bringing about a change, others in preventing a change from occurring; but in either case the notion of change is crucial to an understanding of action, so that, according to von Wright, it will not be possible to develop a *logic* of action without first constructing a logic of change; and this is what his *T*-calculi were designed to provide.

Von Wright's starting point was the Russellian idea that change can be characterized in terms of the difference in truth-value of the same proposition at different times (or, as Russell himself put it, the difference in truth-value of propositions *about* different times but otherwise identical). A change in respect of a proposition p thus comes about either through p's being first true and then false, or the other way round. It was to express this situation that von Wright introduced the binary connective T such that pTq is true now if and only if p is true now and q is going to be true at some later time. Of course, for many possible choices of p and q, no change is implied by the truth of pTq at any time; but if they are incompatible with one another, then some change must occur in going from a state of the world in which p is true to a state in which q is true; so that in particular the proposition $(\neg p)Tp$ expresses the change involved in p's *coming to be true*.

A curious feature of this set-up is that it is only possible to express *future* changes in this way; this is a consequence of propositions of the form pTq being evaluated for truth or falsity at the earlier of the two times involved in the change. The interval semanticists (notably Humberstone) pointed out that the circumstance of p's being true followed by q's being true is something that obtains not at a single moment of time but over an interval: this was one of Humberstone's arguments for basing the semantics on intervals in the first place.

In Galton (1984), I adopted a different approach, focussing attention not on the sequence '$\neg p$ then p' (which as Humberstone rightly pointed out requires an interval for its realization) but on the event of *transition* from $\neg p$ to p, which is by contrast instantaneous. And I noted that a *past* occurrence of this event can be expressed by the tense-logical formula

$$P*(P\neg p \wedge p)$$

whereas a *future* occurrence of the same event is expressible by

$$F*(\neg p \wedge Fp),$$

where $P*\alpha$ and $F*\alpha$ are defined to be equivalent to $\alpha \vee P\alpha$ and $\alpha \vee F\alpha$ respectively. It was the lack of any common formal element between these two formulae that led me to postulate the necessity of a new kind of temporal logic, *Event Logic*, in which events can receive a uniform characterisation. Event Logic is the subject of my paper 'The Logic of Occurrence' (this volume, Ch. 5).

Von Wright's T-calculi have been quite influential in providing a starting-point for the development of later approaches to the logic of change, although it seems that the T-calculi themselves are now of little more than historical interest. It should be noted, incidentally, that von Wright approached temporal logic as a tenser: his T-formulae contain no temporal references, and thus may have different truth-values at different times.

A stark contrast to von Wright's work is the detenser account of action given by Davidson (1967). This arose out of a concern for *logical form*. Davidson's fundamental aim was to provide a Tarski-style semantic analysis of natural language. In order to do this, he followed Quine in adopting a thoroughgoing program of reducing all natural language constructions to first-order logical formulae.

An apparently insurmountable obstacle to this program was the problem of adverbials: what, asked Davidson, is the first-order logical form of a sentence like

$$Jones\ buttered\ toast\ with\ a\ knife \qquad (9)$$

It would not do to postulate a three-place predicate $b(x, y, z)$ to render 'x buttered y with z', for as Anthony Kenny had pointed out (Kenny, 1963, Ch. 8), this would obscure the logical connections of (9) with such an obviously related sentence as

$$Jones\ buttered\ toast\ at\ midnight \qquad (10)$$

in which 'with a knife' is replaced by a new expression bearing a totally different semantic relation to the whole.

One possibility which has found favour with some researchers is to treat the adverbial phrases 'with a knife' and 'at midnight' as *predicate modifiers*, that is as functions which convert the predicate 'buttered' into new predicates 'buttered with a knife' and 'buttered at midnight' respectively. Davidson rejected this approach on the ground that it would mean going beyond the confines of a first-order logical analysis.

The ingenious solution proposed by Davidson has won a wide following. Davidson postulated that (9) and (10) assert the *existence* of certain events: both assert the existence of an event of Jones's buttering toast, but (9) asserts that such an event came about through the use of a knife, whereas (10) merely asserts that it took place at midnight. Thus Davidsonian logical analyses of (9) and (10) look like

$$(\exists x)[\text{butter}(\text{Jones, toast}, x) \wedge \text{with}(x, \text{knife})]$$

and

$$(\exists x)[\text{butter}(\text{Jones, toast}, x) \wedge \text{at}(x, \text{midnight})].$$

With these analyses, the logical relationship between (9) and (10) (namely, that they both imply 'Jones buttered toast') becomes an elementary exercise in first-order logic.

Note how thoroughly Davidson here embraces Quine's belief that the logical content of a sentence can only be brought out by casting the sentence in first-order form. In order to do this he has to have terms referring not just to people and things ('Jones', 'toast', 'knife') but also to times ('midnight') and, most significantly, to events. It is beside the point that events never get to acquire names, appearing only as the bound variables of quantifiers, since in a fuller analysis, according to Quine, the same would be true for all kinds of entities, their names being paraphrased away by means of Russell's theory of descriptions.

Analyses in Davidson's style have been taken further, and integrated with some of the insights into aspectual character mentioned in the last section, by

Barry Taylor, whose recent book 'Modes of Occurrence' (Taylor, 1985) provides a full account of his work on the logic of adverbials and related issues to date.

1.6 Reichenbach's Theory of Tenses

Mention must here be made of the work of H. Reichenbach, whose theory of tense (Reichenbach, 1947) has formed the basis of much subsequent work, including some of immediate relevance to this chapter.

To see clearly the gist of Reichenbach's way of looking at tenses, reflect first that a Prior tense such as P involves implicit reference to *two times*, namely the time at which a sentence Pp is to be assigned a truth-value, and a prior time at which p is being said to have been true. By iterating tense-operators, the number of implicit temporal references can be increased without limit, so that, for example, the truth of a sentence of the form $PFPp$ presupposes a time t_1 at which it is true, a time t_2, earlier than t_1, at which FPp is true, a time t_3, later than t_2, at which Pp is true, and finally a time t_4, earlier than t_3, at which p is true.

Now the essence of Reichenbach's theory is that all the tenses of natural language can be accounted for in a scheme which invokes just *three* times for each tense. Reichenbach calls these times U (for *utterance time*), R (for *reference time*), and E (for *event time*). On this scheme, the future perfect tense seen in 'I shall have gone' can be analysed as presenting us with a set-up in which U (the time at which the sentence is uttered) precedes R (the time spoken of, at which it is to be true that I have gone), and R follows E (the time at which I actually go), the former relation being signalled by the future element 'shall', the latter by the perfect element 'have + *past participle*'.

A notable success of Reichenbach's method was its analysis of the distinction, in English, between the simple past ('I went') and the perfect ('I have gone'). In both, the event time precedes the utterance time, but whereas the reference time coincides with the event time in the simple past, it coincides with the utterance time in the perfect. Assuming that temporal adverbials attach to the reference time of a sentence, this neatly explains why, for example, we can say

$$I\ did\ it\ yesterday \tag{11}$$

$$I\ did\ it\ today \tag{12}$$

and

$$I\ have\ done\ it\ today \tag{13}$$

but not

$$I\ have\ done\ it\ yesterday, \tag{14}$$

since while a *past* reference time may fall within either yesterday or today, thus explaining the acceptability of (11) and (12), a *present* reference time can only fall within today, explaining the acceptability of (13) but not (14).

Reichenbach's work occupies a somewhat ambiguous position in this admittedly interdisciplinary research area. On the one hand, the striking simplicity of the schematism and its genesis in a work on symbolic logic suggest that the theory ought to be regarded as a part of logic, and thus within the domain of philosophy; on the other hand, the enterprise does seem to embody a substantive thesis about language, in that it explicitly denies that logically possible tense constructs such as the tense logical formula *PFPp* cited earlier will find expression in the grammatical structure of natural language—and indeed it does seem to have been amongst linguists rather than amongst logicians that Reichenbach's theory has proved most influential.

2   TEMPORAL LOGIC IN ARTIFICIAL INTELLIGENCE AND COGNITIVE SCIENCE

One of the areas in which Computer Science has found a use for temporal logic is Artificial Intelligence (AI). Whether the aim here is to duplicate human intelligence or merely to simulate certain of its effects, the important thing is that human intelligence manifests itself both in what we do and in what we say, so that much AI research has been directed towards the tasks of simulating *action* and simulating *language*. For both of these tasks, a proper treatment of time is desirable—in the former case because actions take place in time and require an appreciation of the logical structure of temporal facts for their proper planning and execution, and in the latter because, as noted earlier, so much of natural language is pervaded, both at the lexical and at the grammatical level, by temporal notions.

2.1   Bruce's 'Chronos'

An early attempt at mechanizing part of our understanding of time within an AI context was that of Bruce (1972), who presented a formal model of temporal reference in natural language. Built upon a first-order logical base, the model is essentially a highly general theory of *tense*, in which Reichenbach's idea of a tense as a relation holding between three points in time is generalized to relations of arbitrary complexity on *intervals*, the relations being built up out of a basic set of seven binary relations by logical conjunction. In context, the intervals referenced by a given tense may be understood as doing duty for *events*, an event here being understood as an individual occurrence occupying just the interval associated with it.

Temporal expressions in natural language are mapped onto this abstract theory by means of a rather crude translation procedure whereby each element of temporal discourse is taken as the mark of a single formal relation in the model, complex natural language expressions thus corresponding to complex formal tenses by compounding the translations of the components appropriately. Thus, to formalize a sentence like

John was to have gone

we note that the expression 'was to' corresponds in Bruce's theory to a formal relation which may be briefly represented as 'after before', while the past participle (in 'gone') corresponds to 'after', so that the 'tense' of our example is 'after before after', which in Bruce's first-order notation comes out as

$$\text{after}(s1, s2) \land \text{before}(s2, s3) \land \text{after}(s3, s4).$$

Here $s1$ and $s4$ correspond to Reichenbach's utterance time and event time respectively, while $s2$ and $s3$ are both reference times, there being no limit, in principle, to the number of distinct reference times possible in Bruce's theory.

Bruce used his model as the basis for a question-answering program called 'Chronos', which seems to have performed satisfactorily within the limitations of its method. These limitations derive in part from the over-simple nature of the translation procedure, which does not take account of the facts that, first, there is no easy way of *telling* whether a given natural language expression is functioning in a particular context as a tense-marker or as the bearer of a non-temporal meaning (so that, for example, in some contexts 'He was to go' means 'It was the case that he was later going to go'—a temporal meaning of the tense-logical form *PFp*—but in other contexts the same clause may mean 'He was under an obligation to go' or 'There was an arrangement or plan that he should go'), and, second, related to this point, a given tense-marker may indicate different tenses in different contexts (so that, for instance, Bruce's 'Chronos' would presumably be unable to handle a sentence like:

If the employment situation hasn't improved by the time the prime minister calls the next general election, there will be a change of government

in which the present perfect and simple present tenses of the subordinate clauses have temporal references of the form 'before after' and 'before' rather than the 'after' and 'equal' which Bruce's translation procedure would, if I understand it aright, assign to them).

2.2 Kahn and Gorry's 'Time Specialist'

After 1972, not much appears to have been published on modelling temporal inference in AI until the appearance of a paper entitled 'Mechanizing

Temporal Knowledge' by Kahn and Gorry (1977). In this paper, Kahn and Gorry introduce the idea of a *time specialist*, an autonomous body of problem-solving routines specifically designed to handle temporal matters; and they envisage that, once constructed, the time specialist could 'be placed in the service of a larger problem-solving program to deal with the temporal questions that arise in the domain dealt with by the latter' (Kahn and Gorry 1977, p. 88).

Unlike Bruce, Kahn and Gorry made no attempt to incorporate any form of natural language understanding in their system, the temporal facts to be processed being first translated manually into intuitively rather opaque but formally more manageable LISP-like expressions. The authors acknowledge that our everyday temporal discourse contains a variety of different types of expression, which only with a certain artificiality can all be regimented into a uniform style of analysis. To accommodate this heterogeneity, their time specialist is endowed with the capacity to organize temporal facts in its memory in more than one way. To be specific, they may be organized (i) by dates, (ii) by referring all temporal phenomena to a basic collection of 'special reference events', or (iii) by setting up chains of events ordered by the before-after relation. For a given set of temporal facts one or other of these methods may be the most appropriate; a weakness of the system is that the decision as to which scheme should be used is not automated, but has in every case to be made by the user.

It is debatable how far any of this can be said to involve a temporal *logic*. Indeed, in their reluctance to relate the natural diversity of temporal expressions to a common underlying formalism, Kahn and Gorry would appear to be eschewing the idea of a temporal logic. None the less, the very concept of a time specialist, with its implication of a determination to abstract the temporal element of reasoning and study it in isolation, has a clear affinity with the ideas motivating the study of temporal logic, and it is perhaps for this reason, more than anything else, that Kahn and Gorry's work deserves to be mentioned here.

2.3   McDermott's Temporal Logic

Continuing our survey, we next come to Drew McDermott's paper 'A Temporal Logic for Reasoning about Processes and Plans' (McDermott, 1982). This paper offers a 'naive theory of time', in much the same spirit as Patrick Hayes's 'naive physics' (Hayes, 1978). McDermott aims to provide 'a robust temporal logic to serve as a framework for programs that must deal with time'. Clearly the underlying motivation for this is similar to what led Kahn and Gorry to posit the idea of a time specialist as an appropriate goal for AI research; but with McDermott the *logical* character of the enterprise is made explicit.

Like Bruce, McDermott based his temporal logic on first-order logic; and he makes much of two 'key ideas' which dominate his temporal thinking. One of the key ideas is that in order to model continuous change, it is necessary for the time line to be continuous, so that between any two instants there is a continuum of instants. This is achieved by modelling the time line by the set of real numbers. The other key idea is that the future is indeterminate, and hence can only be modelled by having many possible futures for any point in time. To secure this, McDermott does not endow time itself with a branching structure; rather, what he does is to define what happens, or may happen, in time by means of a partially ordered set of *states*, related to the linear time line by means of an order-preserving date-function from states to instants. A state is, in fact, 'an instantaneous snapshot of the universe'. The indeterminacy of the future now appears as a condition on the partial ordering of states, namely that if two distinct states both precede some common third state, then one of the two must precede the other. This ensures that the partial order only branches into the future, giving the total set of states a tree-like structure. McDermott names the maximal *linear* paths through this structure 'chronicles'; so a chronicle is one possible total world-history.

Certain sets of states are singled out as 'facts'. The idea here is that a fact is something which may be true in some states of the world and false in others; and the set of states in which it is true is taken as characterizing the fact. This way of defining facts commits us to the view that there cannot be distinct facts true in exactly the same states of the world. If we were restricted to a single chronicle, this limitation would be disastrous, but it might be argued that the branching structure of the totality of states provides just enough 'leeway' for the mere pattern of incidence in state-space to individuate facts to the right degree of precision, in much the same way that the 'possible worlds' semanticists seek to show that notions which appear intensional when the terms expressing them are interpreted on a universe corresponding to just the actual world become extensional when the interpretation is spread out across the more spacious canvas of a universe corresponding to the totality of all possible worlds.

A set of states can only count as a fact if it obeys certain formal conditions, in particular McDermott's 'Axiom 9', which in effect rules out the possibility of a fact's changing its truth-value infinitely often over a finite stretch of time. This outlawed situation was described by Prior as a 'fuzz', and by Hamblin (1971) as 'indefinitely fine intermingling'; it is of particular interest to us here because it illustrates how careful one must be, when seeking to axiomatize a particular model of time, to rule out explicitly any undesirable 'pathological' consequences of what might seem intuitively the right set of postulates for the model one has in mind. It is significant, perhaps, that both Hamblin and McDermott seem to have needed external prompting in this matter, for the axiom Hamblin uses to rule out indefinitely fine intermingling was suggested to him by Prior, while McDermott attributes his axiom 9 to Ernie Davis.

In addition of facts, McDermott also discusses *events*, and notes that they are harder to handle than facts. He goes on to reject two ideas about how events should be handled before settling on his own approach. The first idea he rejects is that an event can be identified as a *fact change*, i.e. in terms of the facts which characteristically precede and follow the event; and he correctly rejects this idea on the ground that many events just aren't fact changes, an example being Davidson's 'John ran round the track three times'. The second rejected idea is that an event can be identified with the fact that it is taking place; and McDermott notes that this only works for events which 'consist of some aimless thing happening for a while' (what I call 'atelic events' in Galton 1984, pp. 66–68). Other events (and these in fact constitute the great majority of those that are ever of much interest to us) are inhomogeneous in the sense that there is no one fact such that the event just consists of that fact's obtaining for a while.[2]

In the face of these formidable difficulties, McDermott abandons the attempt to give an *internal* characterization of events in terms of configurations of facts in favour of a wholly *external* characterization in terms of their temporal incidence, identifying an event as the set of intervals on which it happens. Now this move is not without value (indeed, I myself adopt a very similar ploy in 'The Logic of Occurrence', see Ch. 5, this volume) but it is important to be clear about precisely what its value is. It seems to me quite legitimate to set up a *correspondence* between events and the sets of intervals on which they occur, in order to reason about the logic of event occurrence by investigating the formal properties of those sets of intervals; I would question the propriety, however, of *identifying* an event with a set of intervals, as McDermott apparently wants to do.[3]

It is not that such an identification would in principle rule out the possibility of retrieving an internal characterization of the event, for in fact it need not, so long as the branching chronicle-tree is expansive enough to cover every eventuality. If, for example, the event of John's driving from London to Leeds is initially defined as the set of intervals on which this

---

2. More precisely, there is no such fact that can be described independently of its being constitutive of the event, for of course if we take an event such as the one reported in the sentence 'John ran round the track three times' then there *is* a fact, namely the one reported in the sentence 'John is in the process of running round the track three times, and will finish doing so', such that our event *does* simply consist of that fact's obtaining for a while; but this fact cannot be described independently of the event.

3. A similar criticism has been made by Turner (1984).

occurs (in every possible chronicle, not just the 'actual' chronicle), then the internal characterization of this event could be derived as the uniquely most inclusive conjunction of facts true of all the intervals in the set. This would include such facts as that John is in London for a while at the beginning of the event, in Leeds at the end of the event, is driving a car most of the time during the event, and so on. Any 'accidental' features such that, say, the car was blue on every occasion that John *actually* drove from London to Leeds, will be ruled out by the inclusion of non-actual, but possible, chronicles in which the car is a different colour.

Rather, it is the conceptual ordering that seems to me crucial; an event is what it is because of its internal characteristics. The external characteristics may suffice for certain purposes, but can never amount to a definition, and indeed, the complete set of chronicles required to retrieve the internal characterization from the external one could not, in principle, be obtained unless we already had at our disposal the very internal characterization we are seeking to retrieve.

McDermott glibly avoids this issue, with the result that it is not clear just what is supposed to be achieved by the rejection of two internal characterizations and their replacement by the external one; to treat them as comparable in this way is to ignore the fundamental difference in type of the two kinds of characterization, and would seem to betoken a gross confusion as to just what is required of a robust foundation for a temporal—or indeed any other—logic.

2.4 Allen's Theory of Time

Another major figure in the AI approach to temporal logic is J. F. Allen, who in a series of papers[4] appearing over a period of several years has elaborated what he calls a *theory of time* intended to play the part of a foundation for the sorts of temporal reasoning required in AI type applications. Allen's approach differs from McDermott's in a number of important ways, and it is these differences that I shall particularly select for discussion here. A general introduction to Allen's theory, in the context of a detailed comparison with two other recent theories, can be found in Fariba Sadri's contribution to this volume (Ch. 4).

Allen endorses the Kahn and Gorry idea of a time specialist, but differs from them in not being concerned with *dates* (Allen, 1981, p. 222). Rather, he is concerned with temporal *relationships*. Moreover, he rejects instants as the basic unit of temporal reference, preferring to replace them with intervals. His reason for this is that 'the only times we can identify are times of occurrences

---

4. Allen (1981), Allen and Koomen (1983), Allen (1984), Allen and Hayes (1985).

and properties' (Allen, 1984, pp. 127–128); and such times, according to Allen, are always decomposable into subtimes, and thus must be treated as intervals rather than as instants. Allen thus implicitly rules out the notion of an instantaneous event, and this ruling seems to me an important weakness of the interval-based approach. For further discussion of this point, see Galton (1986).

Bruce's set of seven basic binary relations on intervals is expanded to nine in Allen (1981) and to 13 in Allen and Koomen (1983). These relations are themselves interrelated by a set of *transitivity relations*, an example of which is that if i is during j and j precedes k, then i precedes k. Because of the transitivity relations, whenever a new fact is added to a network of interval relations, the consequences of this addition reverberate over the whole network. To compute the full extent of this reverberation would require impractically large amounts of memory, so Allen restricts the 'propagation' of relations to pairs of intervals that share a common *reference interval*, an idea suggested by Kahn and Gorry's 'special reference events'. For example, if i has reference interval 1984 and j has reference interval 1985, then on adding the latter fact there is no need to add also the consequence that i is before j, since this is easily computed, when needed, from the three facts 'i during 1984', 'j during 1985', and '1984 before 1985' (See Allen, 1981, pp. 223–224).

So far, we have only discussed the treatment of time itself in Allen's system. This would be of little interest for AI without the superstructure which is built upon it, the theory of what happens or obtains in time. Allen uses a threefold ontology of *properties*, *processes*, and *events*; these may be distinguished by the way in which they hold or occur in time. Thus, a property holds over every subinterval of any interval over which it holds (e.g. if I was in Norfolk all last week, then I was in Norfolk all of last Tuesday). For an event, on the other hand, each occurrence defines a unique interval over which it occurs, and it does not occur over *any* subinterval of that interval. Allen notes the affinity of this treatment of events with McDermott's, but he does not seem to commit the same error of seeking to identify an event with the set of intervals over which it occurs.

Allen's processes sit rather uneasily between properties and events: according to Allen, if I was walking throughout the last hour (Allen says 'over' instead of 'throughout', but it is hard to see what difference this is supposed to make, apart from lending superficial plausibility to what he goes on to claim), then the hour must contain *some* subintervals throughout which I was walking, but it need not be the case that I was walking throughout *every* such subinterval. Unfortunately for Allen's theory, it is surely the case that if there is any time during the hour when I was *not* walking, then it is simply untrue that I *was* walking throughout the hour; and if one interprets 'walking' broadly so that one can be walking (in this broad sense) throughout a period containing intermittent stretches of not walking (in the narrower sense), then in the broad sense of walking one *is* walking even during those stretches (in rather the same way that one can be writing a book even though right now one is having a meal, or asleep). For this reason I am unhappy about processes; but I seem to be pretty much alone in this, since most writers on such matters include this category, or something very like it, in their classifications.

Allen uses three different predicates to relate elements from his three ontological categories to the times over which they hold or occur. For a property, he uses the predicate HOLDS(p, t) to state that property p obtains throughout the interval t; for an event he says OCCUR(e, t); and for a process, OCCURRING(p, t). Allen states axioms for each of these predicates to secure their correct logical behaviour. The axioms for HOLDS turn out to be identical with the axioms given by C. L. Hamblin (1971). As I have argued elsewhere (Galton, 1986), these axioms have the weakness that they make it impossible to give a correct treatment of continuous change. This was singled out by McDermott as one of the three most important problems of temporal representation, to which McDermott's own solution was, as already described, to use a continuum of instants as the basis for temporal reference.

Another important divergence from McDermott's system is that Allen rejects the use of a branching future. His ground for this is that 'reasoning about the future is ... just one instance of hypothetical reasoning' (Allen, 1984, p. 131). Reasoning about the past is also hypothetical in this way, so Allen would as soon introduce a branching past into his model as a branching future. Note the contrast between the kinds of reason given by McDermott and by Allen for their respective decisions in this matter. Allen is concerned with the structure of our reasoning about time, and argues that in this respect the past and the future are on the same footing. McDermott, on the other hand, is concerned with the way things really are: he is impressed by the thought that the future really is indeterminate whereas what is done is done and cannot be undone. It seems to me that Allen's more pragmatic approach is more appropriate to the kind of enterprise we are considering here; McDermott is in danger of getting himself entangled in a mire of philosophical perplexities—after all, philosophers have argued since the time of Aristotle about the nature of the indeterminacy of the future, and are arguably little nearer to achieving a solution acceptable to all.

One of the major goals of both McDermott and Allen's researches on the representation of temporal reasoning is to facilitate automated *planning*, particularly in the domain of problem-solving, this being a hallmark of intelligent human activity and thus a prime target for AI. Allen and Koomen (1983) describe a concept of planning whereby a system is presented with initial and goal states and uses these to construct a partial simulation of the desired future; the planner then looks for 'causal gaps' in the simulation and attempts to fill them by the introduction of *actions*, this procedure being executed recursively until a complete causally connected description of the desired future is achieved.

2.5 Other Recent Approaches

Kowalski and Sergot (1986) have developed a temporal formalism broadly similar to that of Allen, intended for the updating of databases and for understanding narratives. The guiding principle is that since deletion from a database necessarily involves loss of information, it is no good trying to handle change by substituting new facts for old, for then no record will remain of the state of affairs before the change; yet often one needs to know not just how things are now, but also how they were in the past. It is therefore necessary to represent change in the database by explicitly tagging facts with the times at which they are true, and adding new facts as the occasion arises. No more need be said here about the Event Calculus, as Kowalski and Sergot's system is called, since a detailed treatment is given by Fariba Sadri in Chapter 4 of this volume.

Fariba Sadri also discusses the system of Lee *et al.* (1985), which employs formalisms adapted from Rescher and Urquhart and from von Wright. Rescher and Urquhart (1971) introduced a notation which can be regarded as a compromise between, on the one hand, the modal style of tense-logical formulae like Fp and Pp and, on the other hand, first-order formulae of the form $p(t)$, where t stands for a time. Rescher's notation is

$$R_t(p)$$

for 'p is realised at t'. This resembles the first-order notation in explicitly mentioning t, but agrees with the modal approach in respecting the integrity of the proposition p, which may accordingly take any of the forms available within the modal notation, yielding formulae like, e.g. $R_t(Pp)$. In particular, there is no reason why one should not bring in formulae of the kind found in von Wright's T-calculi, e.g. pTq, giving us

$$R_t(pTq).$$

Although Rescher himself does not do this, formulae of this type are to be found in Lee *et al.*'s paper, their notation being $R(t):(p!q)$. To the extent that Rescher's notation is partly modal, partly first-order, the same may be said of Lee·*et al.*'s; but the overall impression I get, having worked mainly with modal temporal logics, is that this formalism has a strongly first-order flavour to it, and indeed Fariba Sadri shows that many of the apparent differences between this system and those of Kowalski & Sergot and Allen do not lie more than skin-deep.

To summarize, it is clear that a logical treatment of time has an important part to play in AI, both from the point of view of simulating natural language understanding and from the point of view of simulating intelligent planning. In addition, similar ideas occur in work on databases, which lies outside the sphere of AI proper. Most workers in the field have chosen a first-order formalism, in which times and, in some cases, events are designated by individual terms. It has become apparent that the depth of the conceptual analyses required to bring this endeavour to a successful conclusion is such as to impinge on many of the traditional philosophical problems concerning time, free will, action, etc.; undoubtedly, the computer science community can benefit from the resulting contact with a hitherto rather remote discipline, and equally, as Aaron Sloman has argued quite persuasively, philosophy itself cannot afford to ignore the new computational developments.

3 THE USE OF TEMPORAL LOGIC IN SOFTWARE ENGINEERING

The process of constructing computer programs can be seen as including at least the following three distinct steps:

(1) Laying down precisely what the program is to do;
(2) Actually writing the program, with the intention that it should satisfy what is laid down in step 1;
(3) Checking that the program really does satisfy 1.

These steps may be referred to as *specification, synthesis,* and *validation* respectively. As we shall see, temporal logic has been applied to all three steps, beginning with validation. Methods of validating programs may be divided into two distinct types, which we may refer to as *testing* and *verification.* Testing a program is a purely empirical matter of actually running it to see whether it works; verification, by contrast, is grounded in theory, and involves mathematically proving that a program is correct. It is to verification rather than testing that temporal logic can be applied.

Before we discuss this in more detail, it is as well to note that, ideally, verification should eliminate the need for testing, but in practice this cannot happen. When one is dealing with a large system, the formal verification becomes exceedingly detailed and cumbersome, and then one's confidence in the correctness of the proofs may sink to a level comparable to the confidence one might have, prior to any verification or testing, that the program itself has been constructed so as to meet its specifications. In any case, even with a relatively small program, no computer programmer would totally refrain from any empirical tests on the ground that the program had already been certified correct by theoretical means. So the two approaches have rather come to be seen as complementary.

3.1 Program Verification

How can one ever be sure that a computer program will always do what its designers intended it to do? This is a problem of great practical urgency as well as considerable theoretical interest. In practice, a program is often tested by being 'put through its paces', i.e. it is run a number of times with a range of different initial conditions selected so as to take into account all the possible sources of error; if it performs these tests satisfactorily (and of course the stringency of the requirements imposed on the program during testing will depend very much on the likely cost of failure in actual use), then it will be given the seal of approval and let loose upon the world, for better or for worse.

This empirical approach to program certification has the drawback that its credibility depends entirely on the testers' confidence that they have indeed foreseen all the sorts of ways in which the program might go wrong, and have devised tests accordingly. Since the number of different possible inputs to a program will, in general, be unbounded, while empirical tests can only sample a small finite portion of these, it is not surprising that there has grown up a demand for a more watertight way of validating programs than empirical testing. This demand has led to highly complex theoretical investigations into the concept of program correctness, and it is in this context that temporal logic has made its entry into theoretical computer science.

3.1.1 Floyd's inductive assertion method

In the early seventies, the dominant paradigm in the theory of program verification was the so-called *inductive assertion method*, due to Floyd, whose paper 'Assigning Meaning to Programs' (Floyd, 1967) is a landmark in the history of this subject. Floyd considered the problem of specifying the behaviour of some simple flowcharts with a view to proving them correct with respect to a given specification. In the following brief account of Floyd's technique, I shall replace flowcharts by programs with assignments and **goto**-statements. Thus adapted, the method consists in assigning to each point in a program a propositional *tag* (known as an *invariant*), in such a way that it can be proved that if the tag at a given point is true at a time when the program's control reaches it, then after execution of the command at that point, the tag at the next point reached will also be true. If, now, the tag assigned to the *start* of the program is true when the program is started up (and this tag can be so devised as to be true so long as the *input specification* of the program is met), then a simple inductive argument shows that if the program halts then the tag assigned to the *end* of the program will then be true—so that if *this* tag is so devised that its truth implies satisfaction of the program's *output specification* then it will have been proved that the correct relation holds between the input and output specifications, i.e. that the program is *correct* (strictly, *partially correct*—see below).

To illustrate with an extremely simple example, consider the following program, which computes factorials:

L_0: **start**

$L_1: f := 1$

L_2: **if** $n = 0$ **then goto** L_6

$L_3: f := n * f$

$L_4: n := n - 1$

L_5: **goto** L_2

L_6: **end**

The tags T_i which are assigned to each program location L_i by the inductive assertion method are as follows:

$T_0: N \in \mathbb{N} \wedge n = N$

$T_1: N \in \mathbb{N} \wedge n = N$

$T_2: N \in \mathbb{N} \wedge 0 \leqslant n \leqslant N \wedge f = N!/n!$

$T_3: N \in \mathbb{N} \wedge 0 < n \leqslant N \wedge f = N!/n!$

$T_4: N \in \mathbb{N} \wedge 0 < n \leqslant N \wedge f = N!/(n - 1)!$

$T_5: N \in \mathbb{N} \wedge 0 \leqslant n < N \wedge f = N!/n!$

$T_6: N \in \mathbb{N} \wedge n = 0 \wedge f = N!$

This example illustrates the three key programming constructs covered by Floyd's account: conditional branching, join of control, and assignment. At a conditional branch of the form

L_i:**if** ϕ **then goto** L_j **else goto** L_k,

the following relations must hold between the tags:

$$T_j \equiv T_i \wedge \phi$$
$$T_k \equiv T_i \wedge \neg \phi.$$

It is easily seen that these are indeed the relations holding amongst T_2, T_6, and T_3 in our example.

Again, where two control-paths merge, the rule is that if L_i and L_j both lead to L_k, then

$$T_k \equiv T_i \vee T_j.$$

In our example, this is precisely the relation holding between T_1, T_5, and T_2.

Finally, the rule for tagging assignment statements is a little more complicated when stated in generality, but easy enough to follow in our example. The assignment $f := n * f$ at L_3 clearly converts T_3 to T_4, since

$$n \times N!/n! = N!/(n-1)!$$

while the next assignment $n := n - 1$ affects both conjuncts of the tag in an obvious way.

That the set of tags shown above constitutes a formal verification of our factorial program follows from the fact that *the tags have been constructed in accordance with the correct rules governing tagging for the relevant program constructs*; what has been proved is that when the program is run with input $n = n_0$, where n_0 is a positive integer, then *if* it ever reaches the terminal location L_6 the values of the program variables at that point will be $n = 0$ and $f = n_0!$. And since the program is intended to compute factorials, we can thus assert that *if* it terminates, it will do so with the intended result.

What the inductive assertion method does *not* prove is that the program *will* terminate. To prove termination, Floyd had to use a completely different method, whereby the program locations are now tagged, not with assertions, but with functions of the program variables whose values are taken from some well-ordered set (typically, the set of natural numbers, or the Cartesian product of this set with itself some number of times). The tagging must be carried out in such a way that during execution of the program the value of the tag, with the current state of the program variables as arguments, steadily decreases with respect to the well-ordering. Given this condition, any possible execution-sequence of the program will correspond to a strictly descending sequence of terms from the well-ordering; from the definition of a well-ordering, any such sequence must be of finite length, and hence so must all execution-sequences of the program, i.e. the program must terminate.

3.1.2 Hoare's axiomatization of the inductive assertion method

The chief importance of Floyd's paper lies in the introduction of a set of *rules* governing the construction of invariants (tags). Hoare (1969) further systematized this idea by formulating such a set of rules as a rigorous axiomatic system, thereby importing the methods of formal logic into the enterprise of reasoning about programs.

Hoare adopted a basic schema of the form '$\{P\}S\{Q\}$', to be interpreted as meaning that if the assertion P is true when the program S is initiated, then the assertion Q will be true if and when it terminates. Hoare used this formalism to state rules specifying the desired behaviour of each construct in his programming language. In his initial paper, the constructs treated are just assignment, sequential composition, and the **while**-statement.

Hoare's system, like any other formal calculus, consists of a set of *axioms* together with a set of *rules of inference* for deriving theorems from the axioms. In fact, Hoare uses only a single axiom, the *Axiom of Assigment*:

$$\vdash \{P[a/x]\}\ x := a\ \{P\} \tag{Ass}$$

Here '$P[a/x]$' denotes the proposition obtained from P by replacing each occurrence of 'x' in P by 'a'. In addition, there are the following rules of inference:

(i) *Rules of Consequence*

$$\text{If } \vdash\{P\}S\{Q\}, \text{ and } Q \text{ implies } R, \text{ then } \vdash\{P\}S\{R\} \tag{Cons 1}$$

$$\text{If } P \text{ implies } Q, \text{ and } \vdash\{Q\}S\{R\}, \text{ then } \vdash\{P\}S\{R\} \tag{Cons 2}$$

(ii) *Rule of Composition*

$$\text{If } \vdash\{P\}S_1\{Q\} \text{ and } \vdash\{Q\}S_2\{R\}, \text{ then } \vdash\{P\}S_1; S_2\{R\} \tag{Comp}$$

(iii) *Rule of Iteration*

$$\text{If } \vdash\{P \wedge B\}S\{P\} \text{ then } \vdash\{P\} \text{ while } B \text{ do } S\ \{\neg B \wedge P\} \tag{It}$$

To illustrate how Hoare's system works, we cannot use exactly the same program as we used to illustrate Floyd's method, because Hoare's logic does not deal with **goto**-statements. Instead, we shall paraphrase our factorial program as a **while**-program, as follows:

> **begin**
>
> $f := 1;$
>
> **while** $n > 0$ **do**
>
> > **begin**
> >
> > $f := n * f;$
> >
> > $n := n - 1;$
> >
> > **end;**
>
> **end.**

Let this program be S; and let the block within the **while**-loop, i.e. '**begin** $f := n*f; n := n - 1;$ **end**', be S_1.

A formal proof of the partial correctness of S can now be written thus:

1. $\{n = N \wedge 1 = 1\} f := 1 \{n = N \wedge f = 1\}$ [Ass]

2. $\{n = N\} f := 1 \{n = N \wedge f = 1\}$ [1, Cons 2]

3. $\{0 < n \leqslant N \wedge n*f = N!/(n - 1)!\} f := n*f \{0 < n \leqslant N \wedge f = N!/(n - 1)!\}$ [Ass]

4. $\{0 < n \leqslant N \wedge f = N!/n!\} f := n*f \{0 < n \leqslant N \wedge f = N!/(n - 1)!\}$ [3, Cons 2]

5. $\{0 \leqslant n - 1 < N \wedge f = N!/(n - 1)!\} n := n - 1 \{0 \leqslant n < N \wedge f = N!/n!\}$ [Ass]

6. $\{0 < n \leqslant N \wedge f = N!/(n - 1)!\} n := n - 1 \{0 \leqslant n < N \wedge f = N!/n!\}$ [5, Cons 2]

7. $\{0 < n \leqslant N \wedge f = N!/n!\} S_1 \{0 \leqslant n < N \wedge f = N!/n!\}$ [4, 6, Comp]

8. $\{0 < n \leqslant N \wedge f = N!/n! \wedge n > 0\} S_1 \{0 \leqslant n < N \wedge f = N!/n!\}$ [7, Cons 2]

9. $\{0 < n \leqslant N \wedge f = N!/n!\}$ **while** $n > 0$ **do** $S_1 \{\neg(n > 0) \wedge 0 \leqslant n < N \wedge f = N!/n!\}$ [8, It]

10. $\{0 < n \leqslant N \wedge f = N!/n!\}$ **while** $n > 0$ **do** $S_1 \{n = 0 \wedge f = N!\}$ [9, Cons 1]

11. $\{n = N \wedge f = 1\}$ **while** $n > 0$ **do** $S_1 \{n = 0 \wedge f = N!\}$ [10, Cons 2]

12. $\{n = N\} S \{n = 0 \wedge f = N!\}$ [2, 11, Comp]

13. $\{n = N\} S \{f = N!\}$ [12, Cons 1]

Note that in this proof every assertion enclosed within braces ought to have an additional conjunct '$n \in \mathbb{N}$', which I have omitted for the sake of clarity. These extra conjuncts ensure that the rules Cons 1 and Cons 2 work correctly: for example, in passing from line 5 to line 6, we make use of the implication:

$$0 \leqslant n - 1 < N \Rightarrow 0 < n \leqslant N.$$

The validity of this implication can only be guaranteed so long as 1, n, and N belong to a data type for which appropriate logical rules hold. The set $\mathbb{N}$ of natural numbers constitutes one such data type, being, of course, the domain of the intended interpretation of the program. A proper axiomatic specification of the data types used by a program is thus an essential part of Hoare's method.

Hoare's paper, though brief and lacking in detail, has been enormously influential. Hoare himself remarked that we should accept the axiomatic proof theory for a programming language as 'the ultimately definitive specification of the meaning of the language'. The idea that the proof system gives the *meaning* of a programming language, and hence of programs themselves, was already explicit in the title of Floyd's paper, and has led to this whole enterprise becoming known as *axiomatic semantics.*

Let us note in passing that in this use of the term 'semantics' there is a curious inversion of the usual use of that term in logic. Ordinarily, the formal semantics of a *proof* system is given by relating it to a *model theory* defined in set-theoretic terms. In the present instance, a proof system (Hoare's logic) is used to specify the semantics of something which is *not* a proof system, namely a programming language (or, more precisely, a class of implementations thereof). Even so, there is still present in this use of the term 'semantics' the all-important idea of a systematic *correspondence* between the entities whose intended meanings are to be formally specified and some other class of entities whose formal properties can be independently given.

This being so, the usual metatheoretical questions regarding soundness and completeness of the axiomatization arise in connection with programming language semantics too, and in particular, if the axiomatic semantics is to be of practical use to actual program-users, it is obviously important to determine precisely *what* class of language implementations a given Hoare-style axiom system is applicable to (cf. Bergstra and Tucker, 1984). Otherwise, for example, one could not be sure that a program which has been certified as correct when run on one implementation of its language will still be correct when run on another implementation.

As originally presented, Hoare's logic dealt with only a very limited set of programming language constructs. Over the years following its introduction, Hoare's ideas have been extended by numerous workers to cover additional constructs such as procedures, recursion, and **goto**-statements (see Apt, 1981, for a survey, and de Bakker, 1980, for a rigorous treatment of many different constructs).

Of particular importance for our present concerns is the attempt by S. Owicki and D. Gries (1976) to adapt the Floyd-Hoare technique to reasoning about *parallel* or *concurrent* computation. With hindsight, it must be admitted that the main conclusion to be drawn from this work is that Hoare's logic is not the most suitable tool for specifying the meanings of parallel programs, and one outcome of this is a proliferation of attempts to find alternative, superior methods (see Barringer, 1985, for a survey).

It is against this background that the temporal approach to program verification to be discussed in the ensuing sections must be understood. This development is a natural consequence of the vastly increased emphasis on the problem of theoretical program verification initiated by the Floyd-Hoare approach, and on the use of logical tools for its solution.

3.1.3   A new approach: Burstall's intermittent assertion method

There is something displeasing about having to use two utterly different methods, one to prove that a program terminates, the other to prove that if it terminates it does so correctly. The latter condition is known as *partial correctness*; the two conditions together constitute *total correctness*. Burstall (1974) devised a method whereby total correctness could be secured in a single proof. Burstall described his method as 'hand simulation with a little induction'. The idea is that one follows through the execution of the program by hand, using symbolic data instead of actual numbers, invoking mathematical induction to prove general statements about what happens at a loop point. As Burstall put it, the prover acts as 'a sort of symbolic interpreter, with a "state vector" giving symbolic expressions instead of numbers as identifier values' (Burstall, 1974, p. 308).

To illustrate the method, let us turn again to our original factorial program. The statement to be proved, i.e. total correctness, can be expressed as

$$\text{For all } N \in \mathbb{N}: \text{ at } L_0 \Rightarrow \text{ sometime (at } L_6 \wedge f = N!).$$

To begin the proof, we assume that the program is at some time at L_0 with $n = N \in \mathbb{N}$. From L_0, control can only pass through L_1 to L_2; after the assignment at L_1 we thus have

$$\text{at } L_2 \wedge n = N \wedge f = 1 \tag{15}$$

We now reach a loop point, so we must use induction. What we prove is that for all i with $0 \leqslant i \leqslant N$, we eventually have

$$\text{at } L_2 \wedge n = N - i \wedge f = N!/n! \tag{16}$$

The base case is straightforward: it is (15) above. For the induction step, assume $0 \leqslant i < N$ and (16). Since $n = N - i > 0$, control must next pass through L_3, L_4, L_5 and back to L_2. Computing the effect of the assignments at L_3 and L_4 gives us

$$\text{at } L_2 \wedge n = N - (i + 1) \wedge f = N!/n!$$

which is the inductive hypothesis for $i + 1$, as required.

Finally, for $i = N$, we have

$$\text{at } L_2 \wedge n = 0 \wedge f = N!$$

and since $n = 0$ the next step is L_6, giving us

$$\text{at } L_6 \wedge f = N!$$

as required.

It is important to note that what we have proved here is not just a conditional, that *if* we get to L_6, then $f = N!$, but the categorical assertion that we *will* get to L_6 with $f = N!$.

Burstall's paper demonstrated the new method on several programs, most of them considerably more complex than our example. The final section of the paper, entitled 'A connection with Modal Logic', is the most significant for our present concerns, for in it Burstall contrasts the implicit form of the statements used in his own proofs, namely

$$\text{sometime (at } L \wedge \ldots),$$

with the form of statement implicit in Floyd's assertions, namely

$$\text{always (at } L \rightarrow \ldots),$$

and notes that both types of statement can be seen as belonging to a simple modal logic. This appears to be the first suggestion that a form of modal logic might be useful in reasoning about programs. Burstall ends with the telling remark that 'further investigation ... might be profitable, asking "what kind of modal logic underlies our informal arguments about program executions?".'

3.2   Further Developments of the Modal Approach

Manna and Waldinger (1978) characterized the Floyd-type assertions of the form 'always(at $L \rightarrow \phi$)' as *invariant assertions*, and the Burstall-type ones of the form 'sometime(at $L \wedge \phi$)' as *intermittent assertions*. They showed that the intermittent-assertion method is at least as powerful as, and never less simple to use than the invariant-assertion method. In particular, they showed how an invariant-assertion proof can always be converted, in a more or less mechanical way, into an intermittent-assertion proof.

Pnueli (1977) explicitly systematized the modal logic adumbrated by Burstall as a *temporal logic* with future-tense operators $\Diamond$ (read 'sometime' or 'eventually') and $\Box$ (read 'always') corresponding to Prior's F and G respectively. The correspondence is not quite exact, in that Pnueli's operators receive an interpretation in which the present counts as part of the future, so that for $\Box\phi$ to be true now, it is necessary not only that ϕ should be true at all future times, but also that it should be true now; and for $\Diamond\phi$ to be true now, it suffices that ϕ itself should be true now, without necessarily being true at any future time. The semantics appropriate for temporal logic in the context of program verification is of course based on *discrete* time, this corresponding to the discrete structure of the execution-sequence. As a result, Pnueli and many others have found it helpful to include in their language an additional temporal operator $\bigcirc$, such that $\bigcirc \phi$ is true at a given time just if ϕ is true at the *next* time.

The other distinctive feature of Pnueli's system is that it presents a logic of *future* time only. This is because the kind of reasoning about programs that is formalized by means of this logic always works forward from a given program-state to the states which succeed it in the execution-sequence. This too has repercussions on the style of the formal semantics. The Kripke semantics for normal tense logic with past and future operators interprets formulae on structures of the form $\langle T, R, t \rangle$, where T is a set of times ordered by the relation R, and t is a member of T singled out as the time *at which* the formula is to be evaluated. The set T, as ordered by **R**, may be dense or discrete, bounded or unbounded, linear or branching. In Pnueli's system, on the other hand, since the temporal formulae are to be used for reasoning about execution-sequences, the structures over which the semantics interprets them must themselves be formal representations of execution-sequences. This constrains $\langle T, R \rangle$ to be a discrete linear ordering isomorphic to the set of natural numbers; and it is no longer necessary to include a third component, t, in the model structures, since it is always tacitly understood that formulae are evaluated on the first term of an execution-sequence. To evaluate a formula on a later term in a sequence, one considers instead the sequence obtained by chopping off a prefix of the original sequence so that the desired term becomes the first term in the new sequence.

It soon became apparent that the new formalism was a useful tool not just for proving program correctness but also for reasoning about programs generally, and that it was particularly suitable for reasoning about *concurrent* programs (in which several processes are executed in parallel, thus as it were competing for the attention of the central processor and requiring to be correctly coordinated in time) and *cyclic* or *reactive* programs (such as operating systems, which do not terminate but rather maintain a continual interaction with their environment). For these tasks, it is of course important to have good abstract models of concurrency, with the result that the application of temporal logic has been an important stimulus towards the development of such models (see e.g. Manna and Pnueli, 1981).

In the next two sections, we shall review the further development of temporal logic in this area of computer science. We shall begin by considering developments in the formal language and its interpretation (*syntactic* and *semantic* developments), and then take a look at the wider context of its use in reasoning about programs (*pragmatic* developments).

3.3 Syntactic and Semantic Developments

It is not possible to keep syntactic and semantic issues sharply separate; this is because syntactic innovations are generally made with a view to increasing, or at least modifying, the expressive power of a language, and expressive power is not a property of syntax alone, but of syntax in relation to a given semantics.

3.3.1 Linear and branching time

A good example of the interdependence of syntax and semantics is furnished by the divergence that has appeared between systems based on linear-time models and those based on branching time. In order to allow for the possibility of interpretations in which the set of times has a branching structure, Lamport (1980) distinguished between tense operators with the meanings 'sometime' (symbolized $\rightarrow$) and 'not never' ($\Diamond$), the latter being defined in terms of 'always' ($\Box$) as $\neg \Box \neg$.

In Lamport's branching-time structure, $\Box p$ is interpreted to mean that p is true throughout every possible future; $\Diamond p$ accordingly means that p is true at *some* time in *some* possible future; while for $\rightarrow p$ the interpretation is that p is true at some time in *every* possible future. Clearly $\Diamond p$ and $\rightarrow p$ are not equivalent in this interpretation; but if we reduce the number of possible futures at each time to one, thereby effectively reverting to a linear-time model, $\Diamond p$ and $\rightarrow p$ do come out as equivalent.

Now Lamport was particularly concerned to establish the relative merits of the two kinds of interpretation in relation to different purposes. He noted that in proving many properties of concurrent programs, a commonly used type of reasoning makes implicit appeal to the principle that any proposition p is either always false or eventually true; and this principle, according to Lamport, implies an axiom of the form

$$\rightarrow p \vee \Box \neg p \qquad\qquad (17)$$

Since this axiom is valid for all models under the linear-time interpretation, but not under the branching-time interpretation, Lamport concluded that linear time was the more appropriate framework for formalizing this kind of reasoning.

Emerson and Halpern (1983) pointed out the flaw in Lamport's argument: the axiom (17) is only a correct formalization of the principle in question so long as the linear-time interpretation is assumed. In effect, Lamport had assumed that (17) expresses the same principle in both linear and branching time. But in branching time, the relevant version of the principle is that p is always false or eventually true for every possible future taken individually; and this principle cannot be expressed in the language used by Lamport.

The required extension to the language was in fact not yet available in 1980, when Lamport's paper appeared. By the time of Emerson and Halpern's paper, there was already in existence a series of branching-time logics of varying expressive power, including some in which it was indeed possible to express the desired version of Lamport's principle.

The crucial innovation underlying the development of this series was due to Ben-Ari *et al.* (1981). They developed a language UB (for '*unified* system of *branching* time') in which the temporal modalities combine quantification over possible futures with quantification over individual times within a future. This gives rise to six modalities, which may be written (here diverging slightly from Ben-Ari *et al.*) as $\forall\Box$, $\forall\Diamond$, $\forall\bigcirc$, $\exists\Box$, $\exists\Diamond$, and $\exists\bigcirc$. The interpretation of these symbols should be clear from, e.g.,

$\exists\Box p$ is true at t iff p is true thoughout some possible future of t;

$\forall\bigcirc p$ is true at t iff p is true at the next time in every possible future of t.

In UB, these symbols are not decomposable: that is, $\forall$ and $\exists$ can only occur immediately preceding $\Box$, $\Diamond$, or $\bigcirc$; and members of the latter set can only occur immediately following $\forall$ or $\exists$.

The branching-time version of Lamport's principle cannot be expressed in UB; but it *can* be expressed if we allow the path-quantifiers $\forall$ and $\exists$ to be prefixed to assertions composed of arbitrary combinations of tense-operators. The required formula is then:

$$\forall(\Diamond p \lor \Box\neg p) \tag{18}$$

In effect, this says that (17) holds in every possible future, as required.

Emerson and Halpern discuss a range of branching-time languages, and nicely characterize their syntactic structures by drawing a distinction between *state formulae*, which describe what is the case at individual times, and *path formulae*, which describe what holds over an entire future path. Our example (18) above is a state formula since an assertion about all possible futures from a given time is an assertion about that time rather than about any one possible future; but what is asserted at each of these futures, namely (17), *is* an assertion about a single entire future, and thus a path formula. So we see that $\forall$ has the effect of converting path formulae to state formulae. On the other hand, $\Diamond$ and $\Box$ convert state *or* path formulae into path formulae.

Emerson and Halpern give a list of eleven such conversion rules which together define a language they call CTL*. By selecting appropriate subsets of this set of rules, one can give exact definitions of the other branching-time logics that have been proposed at various times by different authors.

Applying this form of analysis to the system UB discussed above, we see that this system *lacks* the rule that if p and q are path formulae, so is $p \lor q$: in this language, $\Diamond p$ and $\Box \neg p$ are path formulae, so there exist state formulae $\forall \Diamond p$ and $\forall \Box p$; but because the disjunction of path formulae is not again a path formula, $\Diamond p \lor \Box \neg p$ is not a path formula, so there is no state formula $\forall(\Diamond p \lor \Box \neg p)$. Another branching-time logic derivable in this way is CTL (*Computation Tree Logic*) of Emerson and Clarke (1982). CTL is similar to UB, but unlike the latter includes the 'until' operator U in its formalism. It is mentioned here because we shall have cause to refer to it again below.

3.3.2 Fairness and related properties

Another syntactic innovation, which figures in both linear and branching-time systems, is also closely linked with the issue of expressive power. As we noted above, the use of temporal logic in reasoning about programs need not just be confined to proving correctness. In concurrent programs, an important class of properties are those relating to *fairness*, i.e. the requirement that two or more concurrent processes have as it were equal rights of access to the central processor. This requirement needs to be stated explicitly because from the point of view of the model of concurrency used in the temporal framework the choice as to which process is scheduled at each point in the execution is to all intents arbitrary, being determined in practice by extraneous features to do with the specific implementation of the program; hence fairness cannot be guaranteed *a priori* but must be demonstrable as a program property like any other.

Fairness is not, in fact, a single property, but covers a range of different possible requirements. A weak requirement, for instance, is that a process which is continuously active will eventually be scheduled (this property is called *justice* by Lehmann *et al.*, 1981); a stronger requirement is that a process which is active infinitely often will be scheduled; stronger still is that a process which is active just once will be scheduled. These properties have to do with how much prompting by a process the control of the program needs before it will respond by allocating time to it. Because of this, these properties are called *responsiveness* properties by Gabbay *et al.* (1980), who reserve the term 'fairness' for requirements such as that, of two processes, the one that is active sooner will be scheduled sooner (*strict fairness*).

Responsiveness properties can be expressed using the temporal logic based on $\Box$ and $\Diamond$, but for strict fairness the language must be extended. Gabbay *et al.* introduce for this purpose a binary operator U, read 'until', such that vUw is true now just so long as v will be true at all times until some future time when w will be true. This operator is, of course, one of the pair first introduced by Kamp (see above, p. 11), and Kamp's result, that the temporal logic based on this pair of operators is expressively complete with respect to the first-order properties of a certain class of models, carries through to the logic of Gabbay *et al.*, in which only the *future* fragment of the full temporal logic is considered. One of the main aims of their paper is to present a simpler proof of this result for that future fragment alone.

Program properties requiring U for their expression are called *precedence properties* (Manna and Pnueli, 1981), in contrast to *invariance properties*, which just require $\Box$, and *eventuality properties*, which require $\Diamond$ (Lamport, 1980, calls these latter properties *safety* and *liveness* properties respectively).

Subsequently, a weaker version of U, somewhat misleadingly read 'unless', which does not imply eventual realization of the second argument, has also been found useful. This weaker operator has been variously notated; following Barringer (this volume, Ch. 2), we shall use W. The operators U and W are mutually interdefinable by means of the equivalences

$$pWq \equiv pUq \lor \Box p$$

$$pUq \equiv pWq \land \Diamond q.$$

3.3.3 Extended temporal logic

I remarked above that expressive power is a property of *interpreted* languages, i.e. languages endowed with a semantics as well as a syntax. Expressive *completeness* is a relation involving a further term, since to call a language expressively complete is to say that one can express in it all the properties of some given class. So expressive completeness is, properly speaking, a property that an interpreted language holds in relation to a class of properties, these being, of course, properties of its models.

The sense in which Gabbay *et al.* showed temporal logic with *until* to be expressively complete is that it can be used to express all first-order properties of the execution-sequences of programs. But as Wolper (1981) pointed out, not all properties *are* first-order properties; Wolper's example is the property that a certain proposition is true at every nth step of the computation (and possibly at other steps too). This property cannot be expressed in temporal logic with *until*, yet as Wolper remarks, one might wish to use it in reasoning about a program which has to check for the satisfaction of some condition every n execution steps.

To get round this difficulty, Wolper proposed a method of adding new operators to temporal logic in such a way that its expressive power becomes equal to that of any given right-linear grammar. A right-linear grammar consists of a finite set of production rules of the form

$$V_i \to v_1 \ldots v_j V_k$$

where the V_i are non-terminal symbols, and the v_i terminal symbols of the grammar. Expressions generable by these rules are called *regular expressions*. The connection with temporal operators may be illustrated by the grammar

$$V \to q$$

$$V \to pV.$$

This grammar (Wolper's Example 2, p. 343) generates all sequences of the form

$$q, pq, ppq, pppq, ppppq, \ldots,$$

as well as an infinite sequence of ps; and these are precisely the ways in which an execution-sequence must begin in order for it (or its first term, in a point-based semantics) to satisfy the temporal-logic formula pWg. The link between the grammar and the temporal operator W is that the grammar corresponds, in a systematic way which is expounded in full generality by Wolper, to the following axiomatization of the operator:

$$\vdash pWq \to q \vee (p \wedge \bigcirc(pWq))$$

$$\vdash u \wedge \square(u \to q \vee (p \wedge \bigcirc u)) \to pWq.$$

Now the point is that *any* right-linear grammar gives rise by way of such an axiomatization to a new temporal operator. In particular, it is possible to write a grammar which generates all and only the sequences with p true on every nth step and arbitrary terms elsewhere; Wolper's method then automatically converts this grammar into an axiomatization of a new temporal operator, thus enabling one to express the corresponding property of execution-sequences in the resulting temporal logic. So to each right-linear grammar there corresponds an *Extended Temporal Logic* (ETL), which is expressively equivalent to the grammar. Of course, no one such grammar subsumes all the rest, so this method will certainly not give us a logic that is expressively complete with respect to the totality of regular expressions; rather, what Wolper gives us is a way of extending temporal logic to cover specific classes of regular expressions as required.

3.3.4 Interval semantics

All the systems considered so far have a semantics based on either individual times or whole futures. In linear time, these are equivalent, since there is a one-to-one correspondence between times and their (unique) futures: it makes little difference whether we say that a formula like $\square p$ refers to a property of times (viz. that p is true at every moment from a given time on) or to a property of futures (viz. that p is true throughout a given future). In branching-time models, though, there is no longer this straightforward equivalence, since each time has many possible futures; in this case, then, semantics based on individual times and semantics based on whole futures have different parts to play in the process of building up the interpretation of a formula from basic elements, point-based semantics being appropriate for state formulae, future-based for path formulae.

A radical innovation due to Moszkowski (1983) is to base the semantics on *intervals*. The motivation for this is to facilitate reasoning about finite chunks of program behaviour, as distinct from entire execution sequences. This kind of reasoning has also been applied to computer hardware. The innovation at the semantic level naturally gives rise to syntactic novelty too, including for example the operator ';' (read 'chop'), with the interpretation that $p;q$ is true on an interval just so long as that interval can be decomposed into two contiguous subintervals in such a way that p is true on the earlier of the two, and q on the later. There is nothing corresponding to this operator in languages whose semantics is point-based rather than interval-based; indeed, it is hard to see how there could be, since there is no way of decomposing a point into two component points. Further details of Moszkowski's system ITL (*Interval Temporal Logic*) can be found in Roger Hale's contribution to this volume (Chapter 3).

3.4 Pragmatic Developments

Under the heading of pragmatics are included all those features of a language which go beyond the mere formal description of its syntax and semantics, relating rather to the broader context of how the language is used. The distinction is a somewhat artificial one and cannot do full justice to the fact that syntax, semantics and pragmatics interact in ways which make it impossible to draw sharp boundaries; but taking them as approximate general headings it is clear enough what sort of features are to be included under each.

To help us understand the main pragmatic developments, let us first consider the general question of why temporal logic is such a suitable tool for reasoning about programs. The answer lies in the fact that programming languages and temporal logics can both be thought of as having to do with the description of *sequences*. The operational semantics of a programming language is, in effect, a means by which the set of possible execution-sequences of a program can be derived from the program itself. In parallel with this, the model-theoretic semantics of a temporal logic relates formulae of the logic to the sets of models which satisfy them, and typically these models also take the form of sequences. Identifying the latter sequences with the execution sequences, temporal formulae and computer programs are brought into relation with a common set of structures, and by means of this indirect relationship a direct relation is established between programs and temporal logic formulae, namely that a set of temporal formulae Σ can be said to *specify* a program P just so long as every execution-sequence of P is a model for Σ (see Fig. 1). In this way the programming language can be said to be endowed with a *temporal semantics*.

In the temporal semantics of Manna and Pnueli, the aim is to use temporal logic to prove that a program satisfies certain specifications—namely the statement of its correctness, together with other features like fairness which are involved in the notion of correctness of concurrent programs. The proof takes as its premisses certain temporal formulae which by inspection are obviously satisfied by the program. Temporal logic is then used to deduce other formulae which are less obviously satisfied by the program; that they nonetheless must do so is a result of the pre-established harmony between the model-theory of the temporal language and the operational semantics of the programming language.

As described, this method suffers from a number of serious drawbacks. For one thing, it can only be applied to already-existing complete programs; for another, it generally requires a lot of detailed and tedious working in all but the simplest cases. The developments we shall consider in this section are all aimed at improving this situation.

First, we shall consider attempts to simplify the proof stage by automating the process of deduction. Next, we shall look at attempts to synthesize programs from the temporal logic specifications, thus getting over the drawback inherent in Manna and Pnueli's method that programs must be given in advance. Third, we shall discuss work directed towards *modularizing* the proof process, so that instead of having to be applied to complete

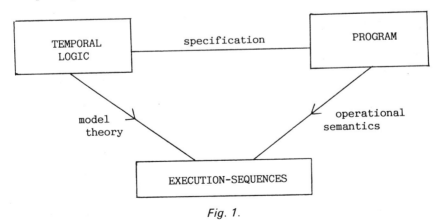

Fig. 1.

programs, temporal reasoning can be applied to smaller program modules in such a way that the reasoning carries over from the separate modules to any larger program of which they are part. Finally, we shall examine attempts to simplify the scheme of Fig. 1 by unifying the temporal specification language with the programming language.

3.4.1 Automatic verification

The goal here is to be able to show automatically that the temporal logic formula expressing a program's correctness is a logical consequence of some set of formulae immediately derivable from the program itself. An obvious way to set about achieving this is to try to construct an automatic *theorem-prover* for the temporal logic in question.

Theorem-proving for ordinary first-order logic generally makes use of *resolution* (Robinson, 1965), so a natural line of inquiry is to look for temporal analogues of the resolution principle. This line has been followed up by Abadi and Manna (1985), who use *non-clausal* resolution, in which the formulae are not first required to be cast into clausal form, and by Fariñas del Cerro (1985), who uses a form of clausal resolution analogous to the clausal resolution principle for first-order logic. These authors have had a certain degree of success in devising workable methods, but it is not yet clear whether it will prove practicable to implement them efficiently for the purposes of automatic program verification.

An alternative approach, due to Clarke *et al.* (1986), is to exploit the model theory of temporal logic rather than its proof theory. The idea is that instead of trying to verify a concurrent system by deducing a formula which expresses the system's correctness from other temporal formulae, we can check directly whether the state-graph of the concurrent system, considered as a Kripke model, satisfies the formula. To do this, of course, requires that the state-graph be finite, so only finite-state systems can be handled in this way; but this is not, the authors claim, too severe a limitation.

Clarke *et al.* use CTL for their specification language (see above, p. 39). The use of a branching-time logic reflects the fact that the totality of execution-sequences, obtainable as the set of possible paths through the state graph, can be viewed as a branching structure. The model-checking algorithm has complexity which is linear in both the size of the specification and the size of the state-graph. This low complexity compares favourably with what could be achieved using a similar method for linear-time temporal logic.

3.4.2 Automatic synthesis

In this section we consider systems that have been devised to synthesize the *synchronization part* of a concurrent program, that is, the part of the program concerned with securing the correct relative timing of the execution of the constituent processes. The synchronization part in most concurrent programs can be handled separately from the *functional part*, which does all the actual computation; and of course it is mainly in reasoning about the synchronization part that temporal logic has proved useful.

Manna and Wolper (1984) use a linear-time propositional temporal logic to specify the individual processes, modifying the resulting specifications in order to reflect the embedding of each process in the environment constituted by all the others, and then they apply a *satisfiability algorithm*, derived from the tableau method of Rescher and Urquhart (1971), to obtain a state-graph for the program. After suitable modifications, the state-graph can be used to derive the synchronization part of the program in the form of usable code.

The model of concurrency used by Manna and Wolper is Hoare's Communicating Sequential Processes (Hoare, 1978), in which a collection of sequential processes interact via input/output operations and in no other way. Emerson and Clarke (1982) use a different model of concurrency, in which the constituent processes can interact by having common access to a shared memory. In order to synthesize the synchronization part (*synchronization skeleton* in their terminology), Emerson and Clarke use CTL rather than linear temporal logic. As with Manna and Wolper, the synthesis proceeds by building up a tableau which can be read as a state-graph for the concurrent system.

Emerson and Clarke defend their use of branching-time logic on the ground that it enables them to assert the *existence* of computation paths having specified properties. This, they say, can be helpful in ensuring that the synthesized program is not a 'degenerate' solution with only a single path. Manna and Wolper, while admitting this, do not regard it as a useful feature, since one is really only ever interested in those properties which hold for *all* computations of a program. Instead, they extol the greater simplicity of linear temporal logic, and the possibility of extending it by the addition of Wolper's 'grammar operators' (see above, p. 41). As things stand, it is not yet clear which of the two approaches will win the day, or if indeed both might have a part to play in the future development of concurrent programs.

3.4.3 Compositional specification

Barringer *et al.* (1984) pointed out that existing temporal logics for program verification suffered from the serious drawback that they could only be applied to complete programs already in existence; whereas what was required for the rigorous systematic development of concurrent programs was a way of using temporal specifications to build up programs bit by bit, specifying and verifying each part, or module, separately, and combining the separate specifications to produce a specification of the whole program: in short, what was required was a *modular* approach, in contrast to the *global* approach prevalent up till then.

In order to do this, Barringer *et al.* have developed a *compositional* system of concurrent program specification using temporal logic. The guiding idea is that the temporal specification language should mirror the language of the program in sufficient detail for the modular structure of the program to be expressible in the specification language. In terms of Fig. 1, what is needed is to make the diagram more nearly symmetrical, with each construct in the programming language (e.g. assignment, concatenation, conditional statement, **while**-statement, parallelism) being given expression in the temporal logic. Explicit appeal to execution-sequences is avoided by suitably axiomatizing the relationship that holds between the programming language and the temporal logic.

To give just one simple example, consider the construct of sequential composition or *concatenation*, whereby two program statements S_1 and S_2 are combined into a new statement $S_1; S_2$. The operational semantics for concatenation asserts that for σ to be an execution-sequence for $S_1; S_2$, σ must be decomposable as $\sigma_1 \circ \sigma_2$, where σ_1 and σ_2 are execution-sequences for S_1 and S_2 respectively. The symbol $\sigma_1 \circ \sigma_2$ here represents the *fusion* of σ_1 and σ_2, the execution-sequence obtained by appending σ_2 to the end of σ_1, if the latter is finite, or else identical to σ_1, if it is infinite. Corresponding to concatenation in the programming language, a new binary *combination* operator C is introduced into the temporal specification language. The semantics of C specifies that an execution-sequence σ satisfies $\phi_1 C \phi_2$ just so long as it can be decomposed as $\sigma_1 \circ \sigma_2$, where σ_1 and σ_2 satisfy ϕ_1 and ϕ_2 respectively.

It is clear from the similarity of the two definitions, of concatenation and combination, that the satisfaction relation thereby set up between program statements and temporal specifications is the right one. The axiomatic definition takes the form of the following rule of inference: if S_1 satisfies ϕ_1, and S_2 satisfies ϕ_2, then $S_1; S_2$ satisfies $\phi_1 C \phi_2$. Thus a correct specification for $S_1; S_2$ can be obtained directly from correct specifications of S_1 and S_2.

In this way, the temporal logic specification of a complex program can be built up from the specifications of its separate components. This technique is still in its infancy; Howard Barringer's contribution to this volume (Ch. 2) presents the current state of thinking in this area.

3.4.4   Temporal logic as a programming language

In all the applications we have considered so far, temporal logic is used as a tool for reasoning about or constructing computer programs written in an existing programming language. Such applications invariably require features of the programming language to be in some way mirrored in the temporal formalism. The clearest illustration of this is perhaps the compositional proof system of Barringer *et al.* discussed in the previous section, in which the compositional structure of programs is reflected by a similar compositionality in the temporal specifications, each programming construct being twinned with its expression in temporal logic.

To some researchers, notably Moszkowski, this twofold formalism has seemed a wasteful duplication, and accordingly attempts have been made to treat temporal logic itself as a programming language. The idea is that instead of first writing a program's specifications in temporal logic and then converting these specifications into another programming language, the program itself should be written in temporal logic. In terms of Fig. 1, this means collapsing the two upper boxes into one, so that the temporal logic formulae become programs, and their relation with the execution-sequences can be simultaneously seen both as a model-theory for the formulae *qua* temporal logic formulae and as an operational semantics for the formulae *qua* computer program. In order to do this, of course, it is necessary to construct an interpreter or compiler for temporal logic.

Moszkowski uses Interval Temporal Logic (ITL), which we discussed above (p. 42); he has adapted a subset of ITL as the programming language *Tempura* (Moszkowski, 1986). Moszkowski himself wrote the first *Tempura* interpreter in Lisp; subsequently, a faster version, written in *C*, has been implemented by Roger Hale. Each *Tempura* program is a formula of ITL.[5] When it is run, an interval satisfying the formula is constructed step by step, each state in the interval corresponding to an action by the computer. The uses to which *Tempura* can be put are described by Roger Hale in Chapter 3 of this volume.

Related work on programming in temporal logic has been done in Japan, where a temporal logic programming language called *Tokio* is being developed (Aoyagi *et al.*, 1985; Fujita *et al.*, 1986). Like *Tempura*, *Tokio* is based on ITL, but it incorporates a somewhat different subset of the full range of ITL formulae. The *Tokio* interpreter is implemented in *Prolog*, and as a result *Tokio* inherits such distinctive *Prolog* features as backtracking and unification, while at the same time allowing a more versatile handling of variables.

In *Prolog*, a variable can be bound to a value only once during execution of a call, so in order to perform operations like incrementing a variable by a

---

5. The converse is not true. This means that the specification language (ITL) is not identical to the programming language (*Tempura*), but rather subsumes it as a special case. The important point is that the relationship between a program and its specification can be defined formally within ITL (Roger Hale, personal communication).

fixed amount it is necessary to introduce an auxiliary variable to assume the resulting value. In *Tokio*, on the other hand, a variable can assume different values at different times (to be precise, a variable gets bound to a *sequence* of values, which can be thought of as the succession of values it assumes over an interval of time), so such operations become as simple to express as in conventional, imperative programming languages.

4 CONCLUDING REMARKS

We have seen application of ideas from temporal logic to two distinct areas of computer science: on the one hand to AI and related domains, and on the other to the theory of programs. These two areas have in fact very little to do with one another; in neither field do researchers ever have cause to refer to what has been done in the other, and the general style of work in the two areas is very different, AI work having a tendency to be discursive and at most semi-formal, while work in the theory of program verification is highly technical and mathematically rigorous. Given this, is there any justification for bringing two such disparate areas together in a single survey?

I believe that a case can be made for the desirability of greater contact between the two fields we have examined here. The very fact that the logic of time is relevant to both is suggestive of the possibility of fruitful interaction. In order to see what possible points of contact there might be, let us make a few comparisons between the two domains.

To begin with, one has to admit that although many of the same issues crop up in both areas of research, they seem to do so in ways which do not provide much common ground for discussion. For example, in both fields the issue of whether to treat time as linear or branching arises; but the criteria that have been brought to bear on this issue in the two cases are utterly different. Ben-Ari *et al.* (1981) stress that in program verification the decision between linear and branching time 'has very little to do with the philosophical question of the structure of physical time', but is rather 'pragmatically based on the choice of the type of programs and properties one wishes to formalise and study'. Contrast with this McDermott's decision to adopt a branching-time model, on the ground that the future really is indeterminate, and Allen's rejection of branching time on the ground that reasoning about the future is just one form of hypothetical reasoning, no different in its essentials from reasoning about the past. Despite this contrast, it remains true that both groups of researchers have to make the choice between the two approaches (and similarly with discrete versus continuous time); so it is not implausible to suggest that a systematic investigation of the formal consequences of each approach would be of assistance in either case.

Another point of contrast is that the work in reasoning about programs is at a far more advanced stage of technical sophistication. The systems of temporal logic used in that work are presented with the highest standards of mathematical rigour, and every attempt is made to provide proofs of consistency and completeness and to compute the complexity of the algorithms employed. The AI work, on the other hand, tends to be much more casual about such matters, with a plethora of axioms and definitions but little attempt to tie these together into a rigorous system which could then be investigated for consistency and completeness (similar criticisms have been made by R. Turner, 1984). It might be argued that this is an unavoidable consequence of the different nature of the tasks involved, AI work being naturally more discursive, even philosophical, in its approach. While this is certainly correct as an analysis of the *cause* of the difference in style, it would be lame if offered as an excuse. If a computer system that employs temporal reasoning is to be used for tasks whose successful execution is important in the real world then it is surely of the utmost importance that the temporal reasoning involved should be mathematically impeccable; otherwise one could have little confidence in the reliability of the results. So there seems to me a good case for importing into AI some of the same mathematical rigour, as regards the logical formalism, that informs the use of temporal logic in reasoning about programs.

A notable difference between the AI and program-verification approaches to time is that in the former, as we have noted, it is customary to use a first-order formalism, whereas in the latter the modal approach is well-nigh universal. AI people have tended to opt for the first-order approach because it forms the basis of existing logic programming facilities, e.g. *Prolog*. Perhaps if it were found possible to automate deduction in modal temporal logics with the same degree of success there would be a drift of opinion in favour of the modal approach, with its greater affinity with the temporal expressions of natural language.

Dov Gabbay's contribution to this volume (Chapter 6) represents an important step towards the realisation of this possibility. In his paper, Gabbay investigates the feasibility of extending Horn Clause logic programming by the inclusion of modal operators. The result, if successfully implemented, would be a tensed version of *Prolog*. Work of this kind is of immediate relevance to the problems of temporal reasoning encountered in AI, and could very likely be of assistance to software engineers as well. In any event, there is clearly much work to be done in this field, and since both the areas we are considering stand to gain by such work there is a clear case for the pooling of resources.

In conclusion, then, it may be said that although the two areas of computer science interested in temporal reasoning have had remarkably little to do with each other, there is enough common ground to allow a certain amount of collaboration between them. In particular, recent developments in the direction of automated deduction in modal temporal logics cannot fail to be of concern to both areas, and should provide a sound basis for further interaction.

Acknowledgements

I am grateful to Roger Hale, Peter Millican, and John Tucker for reading a draft of this paper and for making numerous constructive criticisms of it.

References

Abadi, M. and Manna, Z. (1985), 'Nonclausal temporal deduction', in R. Parikh (ed.) *Logics of Programs*, Lecture Notes in Computer Science (Springer Verlag) **193**, 1–15.

Ackrill, J. L. (1965), 'Aristotle's distinction between *Energeia and Kinesis*', in Bambrough (ed.) *New Essays on Plato and Aristotle* London: Routledge and Kegan Paul, 121–141.

Allen, J. F. (1981), 'An interval-based representation of temporal knowledge', *Proc. 7th Int. Joint Conf. on AI*, 221–226.

Allen, J. F. (1984), 'Towards a general theory of action and time', *Artificial Intelligence*, **23**, 123–154.

Allen, J. F. and Koomen, J. A. (1983), 'Planning using a temporal world model', *Proc. 8th Int. Joint Conf. on AI*, 741–747.

Allen, J. F. and Hayes, P. J. (1985), 'A common-sense theory of time', *Proc. 9th Int. Joint Conf. on AI*, 528–531.

Anscombe, G. E. M. (1964), 'Before and after'. *Philosophical Review*, **73**, 3–24.

Aoyagi, T., Fujita, M. and Moto-Oka, T. (1985), 'Temporal logic programming language *Tokio*: programming in *Tokio*', *Logic Programming '85*, Lecture Notes in Computer Science (Springer), **221**, 128–137.

Apt, K. R. (1981), 'Ten years of Hoare's logic', *ACM Transactions on Programming Languages and Systems*, **3**, 431–483.

de Bakker, J. W. (1980), *The Mathematical Theory of Program Correctness* London: Prentice-Hall.

Barringer, H. (1985), *A Survey of Verification Techniques for Parallel Programs*, Lecture Notes in Computer Science (Springer Verlag), **191**.

Barringer, H., Kuiper, R. and Pnueli, A. (1984), 'Now you may compose temporal logic specifications', *Proc. 16th ACM Symp. on Theory of Computing*, 51–63.

Ben-Ari, M., Pnueli, A. and Manna, Z. (1981), 'The temporal logic of branching time', *Proc. 8th ACM Symp. on Principles of Programming Languages*, 164–176.

van Benthem, J. F. A. K. (1983), *The Logic of Time* Dordrecht: D. Reidel.

Bergstra, J. A. and Tucker, J. V. (1984), 'The axiomatic semantics of programs based on Hoare's logic', *Acta Informatica*, **21**, 293–320.

Bruce, B. C. (1972), 'A model for temporal references and its application in a question-answering program', *Artifical Intelligence*, **3**, 1–25.

Burstall, R. M. (1974),'Program proving as hand simulation with a little induction', *Information Processing*, **74**, 308–312.

Clarke, E. M., Emerson, E. A. and Sistla, A. P. (1986), 'Automatic verification of finite-state concurrent systems using temporal logic specifications', *ACM Transactions on Programming Languages and Systems*, **8**, 244–263.

Cresswell, M. J. (1977), 'Interval semantics and logical words', in C. Rohrer (ed.) *On the Logical Analysis of Tense and Aspect* Tübingen: Gunter Narr.

Davidson, D. (1967), 'The logical form of action sentences', in N. Rescher (ed.), *The Logic of Decision and Action* University of Pittsburgh Press, 81–95.

Dowty, D. (1979), *Word Meaning and Montague Grammar* Dordrecht: D. Reidel.

Emerson, E. A. and Clarke, E. M. (1982), 'Using branching time temporal logic to synthesize synchronization skeletons', *Science of Computer Programming*, **2**, 241–266.

Emerson, E. A. and Halpern, J. Y. (1983), ' "Sometimes" and "not never" revisited: on branching vs. linear time', *Proc. 10th ACM Symp. on Principles of Programming Languages*, 127–140.

Fariñas del Cerro, L. (1985), 'Resolution modal logics', in K. R. Apt (ed.) *Logics and Models of Concurrent Programs* Springer Verlag, 27–56.

Findlay, J. N. (1941), 'Time: a treatment of some puzzles', *Australasian Journal of Philosophy*, **19**, 216–235.

Floyd, R. W. (1967), 'Assigning meaning to programs', *Proceedings of Symposia in Applied Mathematics* (American Mathematical Society), **19**, 19–32.

Fujita, M., Kono, S., Tanaka, H. and Moto-Oka, T. (1986), 'Tokio: logic programming language based on temporal logic and its compilation to Prolog', *Proc. 3rd Int. Conf. on Logic Programming*, Lecture Notes in Computer Science Springer Verlag, **225**, 695–709.

Gabbay, D., Pnueli, A., Shelah, S. and Stavi, J. (1980), 'On the temporal analysis of fairness', *Proc. 7th ACM Symp. on Principles of Programming Languages*, 163–173.

Galton, A. P. (1984), *The Logic of Aspect* Oxford: Clarendon Press.

Galton, A. P. (1986), 'A critical examination of J. F. Allen's theory of action and time', Report 6.86, Centre for Theoretical Computer Science, University of Leeds, England.

Geach, P. T. (1965), 'Some problems about time', *Proceedings of the British Academy*, **51**, 321–336.

Hamblin, C. L. (1971), 'Instants and intervals', *Studium Generale*, **24**, 127–134.

Hayes, P. (1978), 'The naive physics manifesto', in D. Michie (ed.) *Expert Systems in the Microelectronic Age* Edinburgh: Edinburgh University Press.

Hoare, C. A. R. (1969), 'An axiomatic basis for computer programming', *Communications of the ACM*, **12**, 576–583.

Hoare, C. A. R. (1978), 'Communicating sequential processes', *Communications of the ACM*, **21**, 666–677.

Humberstone, I. L. (1979), 'Interval semantics for tense logic', *Journal of Philosophical Logic*, **8**, 171–196.

Kahn, K. and Gorry, G. A. (1977), 'Mechanizing temporal knowledge', *Artificial Intelligence*, **9**, 87–108.

Kamp, J. A. W. (1968), *Tense Logic and the Theory of Linear Order*, Ph.D. thesis, University of California, Los Angeles.

Kanger, S. (1957), 'A note on quantification and modalities', *Theoria*, **23**, 133–4.

Kenny, A. (1963), *Action, Emotion and Will* London: Routledge and Kegan Paul.

Kowalski, R. A. and Sergot, M. J. (1986), 'A logic-based calculus of events', *New Generation Computing*, **4**, 67–95.

Kripke, S. (1963), 'Semantical considerations on modal logic', *Acta Philosophica Fennica*, **16**, 83–94.

Lamport, L. (1980), ' "Sometime" is sometimes "not never": on the temporal logic of programs', *Proc. 7th ACM Symp. on Principles of Programming Languages*, 174–185.

Lee, R. M., Coelho, H. and Cotta, J. C. (1985), 'Temporal inferencing on administrative databases', *Information Systems*, **10**, 197–206.

Lehmann, D., Pnueli, A. and Stavi, J. (1981), 'Impartiality, justice and fairness: the ethics of concurrent termination', *Proc. 8th Coll. on Automata, Languages and Programming*, Lecture Notes in Computer Science, Springer Verlag, **115**, 264–277.

Manna, Z. and Pnueli, A. (1981), 'Verification of concurrent programs: the temporal framework', in R. S. Boyer and J. S. Moore (eds) *The Correctness Problem in Computer Science* London: Academic Press, 215–273.

Manna, Z. and Waldinger, R. (1978), 'Is "sometime" sometimes better than "always"? Intermittent assertions in proving program correctness', *Communications of the ACM*, **21**, 159–172.

Manna, Z. and Wolper, P. (1984), 'Synthesis of communicating processes from temporal logic specifications', *ACM Transactions on Programming Lanuages and Systems*, **6**, 68–93.

Massey, G. (1969), 'Tense logic! Why bother?', *Noûs*, **3**, 17–32.

McArthur, R. P. (1976), *Tense Logic* Dordrecht: D. Reidel.

McDermott, D. (1982), 'A temporal logic for reasoning about processes and plans', *Cognitive Science*, **6**, 101–155.

Moszkowski, B. (1983), *Reasoning about Digital Circuits*, Ph.D. thesis, Stanford University, USA.

Moszkowski, B. (1986), *Executing Temporal Logic Programs* Cambridge University Press.

Newton-Smith, W. H. (1980), *The Structure of Time* London: Routledge and Kegan Paul.

Owicki, S. and Gries, D. (1976), 'An axiomatic proof technique for parallel programs', *Acta Informatica*, **6**, 319–340.

Pnueli, A. (1977), 'The temporal logic of programs', *Proc. 18th IEEE Symp. on Foundations of Computer Science*, 46–67.

Prior, A. N. (1955), 'Diodoran modalities', *Philosophical Quarterly*, **5**, 205–213.

Prior, A. N. (1967), *Past, present and future* Oxford: Clarendon Press.

Quine, W. V. (1965), *Elementary Logic* (revised edition) New York: Harper and Row.

Reichenbach, H. (1947), *Elements of Symbolic Logic* New York: Macmillan.

Rescher, N. and Urquhart, A. (1971), *Temporal Logic* Springer Verlag.

Richards, B. (1982), 'Tense, aspect, and time adverbials', *Linguistics and Philosophy*, **5**, 59–107.

Robinson, J. A. (1965), 'A machine-oriented logic based on the resolution principle', *Journal of the ACM*, **12**, 23–41.

Russell, B. (1903), *Principles of Mathematics* London: George Allen and Unwin.

Sloman, A. (1978), *The Computer Revolution in Philosophy* Hassocks: Harvester Press.

Taylor, B. (1985), *Modes of Occurrence* Oxford: Basil Blackwell.

Tichý, P. (1985), 'Do we need interval semantics?', *Linguistics and Philosophy*, **8**, 263–282.

Turner, R. (1984), *Logics for Artificial Intelligence* Chichester: Ellis-Horwood.

Vendler, Z. (1967), *Linguistics and Philosophy* Ithaca: Cornell University Press.

Wolper, P. (1981), 'Temporal logic can be more expressive', *Proc. 22nd IEEE Symp. on Foundations of Computer Science*, 340–348.

von Wright, G. H. (1965), 'And next', *Acta Philosophica Fennica*, **18**, 293–304.

von Wright, G. H. (1966), 'And then', *Commentationes Physico-Mathematicae of the Finnish Society of Sciences*, **32**, 7.

2 THE USE OF TEMPORAL LOGIC IN THE COMPOSITIONAL SPECIFICATION OF CONCURRENT SYSTEMS

Howard Barringer

*Department of Computer Science,
University of Manchester*

ABSTRACT

This chapter considers the application of temporal logics in the formal specification and development of complex computing systems. In particular, the relevance of compositional proof theories, modularity and abstractness are motivated. A basic technique for obtaining compositional program proof theories, using a temporal language as the assertion language, is demonstrated. Several specialisations for various *real* parallel programming language frameworks are indicated. Finally, a problem in obtaining abstract semantic descriptions of programming languages in the temporal framework is discussed, together with one particular solution suggesting the use of a logic based on a linear dense time model.

1 INTRODUCTION

Over the last 10–15 years, rapid advances in technology have brought computing devices in one form or another to the everyday life of the general public. We find micro-computer devices in such diverse systems as:

pocket calculators, watches, washing machines, microwave ovens, telephones, electronic games, etc.
banking systems, automated office environments, etc.
patient-monitoring systems, etc.
flight control systems, nuclear reactor control systems, etc.
defence (and attack) systems.

Temporal Logics and their Applications

ISBN: 0-12-274060-2

It is abundantly clear that all the above systems are *extremely complex* and very often they *fail*. What happens then? Of course, that depends on the system. One is not too annoyed if some function on a pocket calculator does not match up to the spartan documentation; one just gets frustrated. When water comes flooding out of the front-loading washing machine, owing to some system or program error, the air is much hotter! But after a series of angry letters things usually settle down again. Quite naturally, matters are many times more inconvenient when the bank makes errors, and they always seem to occur in the bank's favour. As for the other types of systems mentioned above, one need say no more than that it is a major catastrophe when they fail. And who is to blame?

Because of the vast size and inherent complexities of these automated control systems employed in safety-critical situations, existing (informal) system development techniques are inadequate. Many scientists *are* concerned and the question 'how do you build large complex systems reliably so that they work as expected?' is taken most seriously and is supported with many well-funded lines of research.

1.1 The Formal Paradigm

One line of research advocates the use of *formal methods* in system development in order to obtain greater confidence and safety in end products. This is not at all surprising. Programming machines is, by its very nature, a most formal task. However, because of the inherent complexities and size of systems, formality in development (design and production) tends to be ignored in favour of, sometimes, just ad hoc methods; methods which appear, on the outset, to be easy to use and understand for the average systems software engineer.

Research over the last 20 years has developed formal techniques for the development of sequential, or rather transformational, systems. Such techniques are now quite well accepted within the research community and are being adopted in many industrial applications. Unfortunately, most of the safety-critical systems mentioned above are of a more reactive nature and cannot be sensibly viewed within the transformational framework. Most often, these are systems where concurrent processing is involved. Although there has been much research in concurrent system specification and verification, as yet there is no general consensus on what are the appropriate techniques for particular types of system. In the following sections of this paper, some insight into the use of temporal logics in programming-language semantics and specification will be given. However, before such exposure is made, it is useful to explore the formal specification and verification paradigm a little deeper.

Consider the initial development stages from the systems life cycle in a formal methods setting. A client produces a set of informal requirements, describing expected behaviour, size, colour, etc. of the desired system. From these, an abstract formal specification is produced. It is abstract in that irrelevant implementation detail is absent, thus providing a clear description of pertinent system properties. From such a formal specification, various refinements, or reifications, are made over many stages until some final 'implementation' is reached. Each refinement step incurs some proof obligation which, when discharged, ensures consistent behaviour of the new development. Assuming each refinement is formally verified, the end result, i.e. 'implementation', will have been proved to have behaviour consistent with the original formal specification, often wrongly referred to as having been proved correct.[1]

In principle, this all sounds fine; however, several criticisms are often raised. For example:

> What guarantee is there that the system behaves according to the client's original expectations? After all, there is a gap between informal requirements and the top-level formal specification, which cannot be linked by formal reasoning.

> As soon as realistically sized systems are considered in this formal paradigm, shortcuts have to be taken; the number of formal proofs required is just far too large. Anti-formalists argue that as soon as such short-cuts are made, gaps in the chain of formal reasoning occur, thus destroying any confidence that might have resulted.

Although the above two points are indeed valid, researchers supporting formal methods argue that it is *rigour* that is required, as employed in a mathematical proof, rather than complete formality. To a large extent confidence in the final product stems from the additional thought, time and precise documentation that ensues from rigorous development. Even so, one is still inclined to ask:

> What constitutes rigorous reasoning?
> How is it distinguished from sloppy formal reasoning?

Such questions are really beyond the scope of this article and are left for the reader to muse upon.

---

1. Other stages in the life-cycle, e.g. testing and maintenance, are still necessary, although both traditional testing effort and post-production maintenance should be significantly reduced.

1.2 Specifications

Clearly, a specification of a system can be given at many levels of abstraction; the deeper the view of the system, the more complex the specification will be. Hierarchic development suggests that development of a system should start at a level of abstraction which is high enough to allow easy expression and understanding of pertinent system behaviour but low enough for the specification not to be trivial. The development process, i.e. the various levels of refinement, adds complexity and detail to the specification by way of adding structure, e.g. decomposing the specification into some particular composition of subspecifications. Specifications are often classified as either *constructive* or *non-constructive*. Constructive specifications provide an abstract model of the desired system; the behavioural properties exhibited by the model are those desired for the specified system. Non-constructive specifications describe such behavioural properties directly. For example, constructively, an unbounded last-in first-out (LIFO) stack might be modelled by an unbounded list, representing the contents of the stack, together with appropriately defined operations *push* and *pop*. A non-constructive specification of the LIFO stack will describe the behaviour purely in terms of the *observable* operations, e.g. by the invariant relations between the operations, for example, for any element e and stack s, $pop(push(e, s)) = s$. In the past, there has been hot debate between research schools supporting these apparently different views of specification. However, it is now largely accepted that both forms have an appropriate place in system development. Furthermore, in certain circumstances it is difficult to differentiate between the two approaches.

When considering concurrent or reactive systems, it is apparent that the behavioural properties fall under two headings. The properties were aptly characterized by Owicki and Lamport (1982) as:

> **safety** informally meaning 'something bad will not happen';
> **liveness** informally meaning 'something good will happen'.

For example,

> *it is always the case that any message delivered by the mail system will not have been phantomly generated*

and

> *the reactor will never go critical*

are obviously safety properties.

For examples of liveness properties consider

all correctly addressed mail will eventually be delivered to its correct destination

and

whenever the reactor is instructed to shutdown, it will eventually do so.

(In the latter case, of course, assuming the safety property of the reactor, it will shutdown *before* it goes critical.)

It should be abundantly clear that a specification language based upon tense logic will provide the right sort of connectives for expressing such safety and liveness properties (in a linear-time temporal framework, $\Box$ is interpreted as 'always in the future', $\Diamond$ is taken as 'eventually' or 'sometime in the future'). Naturally, the difficult question to answer is 'what tense logic?'. Should it be based on classical, relevant, intuitionistic, etc. reasoning? Should time be linear, or branching, or parallel, etc? Should it be discrete or continuous? To a large extent, the structure of time is dictated by the kind of computation model chosen or desired. In the following sections, a logic with a linear time structure is used and is based, primarily, on a discrete time model; however, advantages in basing the logic on a dense time model are exemplified.

1.3   Hierarchic Development and Compositional Specification

The style of development hinted at above, i.e. starting from some abstract level of specification and developing through a series of reifications towards an implementation, is in essence top-down, or hierarchic. Hierarchy in development arises not only quite naturally when one strives for abstract specifications but also when one considers large developments. Consider the following mythical scenario depicted in Fig. 1. A large computer systems manufacturer is developing a new product. The product design team, consisting of a few senior design consultants, develop an initial, high-level specification from the originally given requirements documents. The team decides that the best development can be achieved by using their three crack development teams in New York, London and Tokyo. The design team separate the development, therefore, into three independent tasks by producing three specifications for their crack teams. It really goes without saying that the specifications must be independent so that the crack teams can work, for the most part, in isolation. This requires that the specification 'A', for example, must capture the interactions, or interference, that the implementations of 'B' and 'C' have on 'A'.

Putting these notions into the context of formal program development one

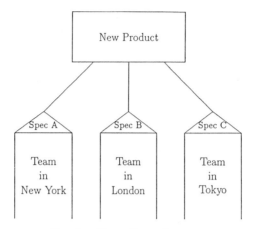

Fig. 1.   Firewall requirements.

should realise that a compositional proof method is required. Informally, a proof method is said to be compositional if the proof rules allow the specification of a program to be verified just in terms of the specifications of the program's immediate syntactic subprograms. Figure 2 displays the way in which specifications can be decomposed towards implementations in a top-down step-wise fashion. With respect to verification, for example, the decomposition of $SPEC_{111}$ into the parallel composition of $Spec_{1111}$, etc.,

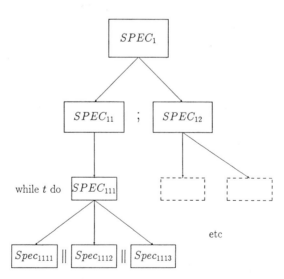

Fig. 2.   Hierarchic development.

must be achieved just in terms of $Spec_{1111}$, etc., without considering any further decompositions.

Obtaining compositional proof methods for sequential programming languages is straightforward. As terminating sequential systems operate, in general, by obtaining input, performing some computation in isolation, and finally delivering outputs, specifications of such can be adequately given as relations between input and output values (relations allow for non-deterministic computation). Such a relational framework admits simple compositional semantics and proof systems for sequential programming languages, see Hoare (1969) and Jones (1980) for example. However such a simple relational framework is not so suitable for the semantics of components that admit interference from a parallel environment. In the seminal papers of Owicki and Gries (1976a, b) that present an axiomatic proof method for parallel programs with fully-shared variables, the proof rules for the parallel composition of program statements are *non-compositional*. The approach uses, however, the concept of meta-level *non-interference* tests that can only be checked when individual components of a parallel composition have, essentially, been fully developed. Jones (1983) was one of the first attempts at developing a compositional proof rule for the parallel composition of statements in a fully shared variable framework. Jones' approach lifts the notion of interference into the specification of a component through the use of rely and guarantee conditions. These conditions capture, respectively, the way in which shared variables are assumed to change under the environment and the way in which shared variables are changed by the component. Unfortunately the expressiveness of such a relational framework is rather limited for specifying general concurrency. The approach presented in the following sections of this paper is, however, based heavily on the notion of separating environment and component actions in specifications. Similar notions also underlie the approach advocated in Lamport (1983a, b).

An excellent account of the 'Quest for Compositionality' is given in de Roever (1985) and sequel papers.

1.4 Organization

The remainder of this chapter is organized as follows. Section 2 introduces a primitive language that acts as a demonstration vehicle for compositional temporal semantic descriptions and compositional proof methods. Also introduced in that section is a discrete linear temporal logic, an ω-regular language, used to describe computation behaviours. Section 3 briefly discusses extension of the techniques to more realistic shared-variable and synchronous message-based parallel programming languages. Section 4 then raises one issue of considerable concern: that of obtaining abstract semantic descriptions. The section indicates how abstract descriptions can result when a temporal language based on a dense time model is employed.

2 COMPOSITIONAL TEMPORAL SPECIFICATIONS

2.1 An Action Language

As indicated through the discussion of the earlier sections, in order to support formal hierarchic program, or system, development, it is necessary to possess a *compositional* proof theory for the implementation language or system. This section of the paper motivates a basic technique for obtaining compositional proof theories for various kinds of 'parallel' programming language, proof theories that use temporal logic as the assertion language. Exemplification starts with a primitive language, referred to as an 'action language'; it is an abstraction of real programming languages. From such a starting point, semantics and proof theories for other, more realistic, languages can be obtained (see Fig. 3, p. 77).

The action language consists of a basic alphabet of actions $\mathscr{A}$, of which a is a typical member ($\in \mathscr{A}$). These actions are atomic, in the sense that they can not be decomposed, at this level of abstraction, or observation. Furthermore, there is an alphabet of 'tests', $\mathscr{T}$, with t as a typical member. Finally, there is an alphabet of procedure names $\mathscr{P}$, with p as typical member.

A program consists of some statement, in an environment of procedure definitions. Statements can be either a call of some procedure defined in the procedural environment, an atomic action followed by some statement (referred to as atomic sequential composition), a guarded choice, or the parallel composition of two statements.

Letting $\mathscr{S}$ denote the class of statements, with S as a typical element, we assume the following syntax for statements.

$$
\begin{array}{lll}
S ::= p & | & \text{procedure call} \\
\quad a \cdot S & | & \text{atomic sequential composition} \\
\quad \square_{k \in \kappa} \, t_k \to S_k & | & \text{guarded choice}[2] \\
\quad S_1 \| S_2 & | & \text{parallel composition}
\end{array}
$$

Two restrictions are placed on the above statement forms in the form of context conditions. Firstly, all recursion must be guarded by at least one

---

2. This form of statement is a generalization of conditional, case and non-deterministic choice. For example, $(t \to S_1 \, \square \, \neg t \to S_2)$ is the standard conditional statement (**if** t **then** S_1 **else** S_2), and (**true** $\to S_1 \, \square$ **true** $\to S_2$) is the (unkind) non-deterministic choice ($S_1 \, \square \, S_2$).

atomic action. This restriction will ensure that no attempts to find solutions to equations of the form $p \stackrel{\text{def}}{=} p$ will be made. Secondly, the sets of actions used by the components of a parallel composition must be disjoint. Thus, letting $\alpha(S)$ denote the set of actions used by S, i.e. defining it as the minimal solution to the following equations,

$$\alpha(p) = \alpha(S_p) \quad \text{where } p \stackrel{\text{def}}{=} S_p$$

$$\alpha(a \cdot S) = \{a\} \cup \alpha(S)$$

$$\alpha(\square_{k \in \kappa} t_k \to S_k) = \bigcup_{k \in \kappa} \alpha(S_k)$$

$$\alpha(S_1 \| S_2) = \alpha(S_1) \cup \alpha(S_2)$$

it is required that for any parallel composition $S_1 \| S_2$, $\alpha(S_1) \cap \alpha(S_2) = \varnothing$.[3]
Programs, denoted by $Pr(\in \mathscr{P}\mathscr{R})$, are then written following

$$Pr ::= \left\{ p \stackrel{\text{def}}{=} S_p; \right\}_{p \in P' \subseteq P} S.$$

Example

Consider the following trivial program,

$$p \stackrel{\text{def}}{=} a \cdot p; \qquad q \stackrel{\text{def}}{=} b \cdot q; \qquad p \| q$$

which consists of the parallel execution of calls of procedure p and q. p will perform an infinite number of a actions, and q will perform an infinite number of b actions. Since the actions of p are disjoint from those of q, there can be no interference between the parallel executions of p and q. It is therefore safe to adopt a strictly interleaved model of parallel execution. The executional behaviour of the example program is thus given by all possible fair interleavings of the behaviours of p and q. Such behaviours can be conveniently written in terms of ω-regular expressions.[4] Thus

$$(b(b^*a)^*)^\omega \cup (a(a^*b)^*)^\omega$$

represents the set of fairly interleaved behaviours of p and q.

---

3. This seems to be quite a strong restriction; indeed it is! Without it, dynamic process creation would be possible, e.g. consider the procedure definition

$$p \stackrel{\text{def}}{=} a \cdot p \| b \cdot p.$$

For ease of exposition, such constructs are handled in the later developments of more realistic languages, see for example Barringer *et al.* (1985).
4. This is the class of expressions of the form $\alpha(\beta)^\omega$, where α and β are regular expressions.

A temporal logic extended with fixed points has proved to be a most useful logic for describing such sets of execution behaviours. This is described below.

2.2   A Temporal Logic with Fixed Points

A propositional temporal language whose models can be viewed as sequences of sets of propositions (i.e. discrete and linear) is presented. The language differs from that given in Manna and Pnueli (1981) in that a notation for solutions to temporal implications and equivalences, i.e. fixed-point constructors, is introduced. The restricted use of such fixed point constructors gives rise to a temporal language equivalent to ETL, Wolper (1983), but in a more natural formalism. For convenience, quantification over propositions is also admitted.

2.2.1   Syntax

The basic symbols of the language are drawn from a alphabet, $\mathscr{A}_L$, of local (state) proposition symbols, an alphabet, $\mathscr{A}_G$, of proposition symbols that appear bound to quantifiers, the normal propositional connectives and quantifiers,

$$\textbf{true false } \neg \vee \wedge \Rightarrow \Leftrightarrow \forall \exists$$

and temporal connectives,

$$unary \begin{cases} \bigcirc ----- \text{next} \\ \square ----- \text{always} \\ \diamond ----- \text{eventually} \end{cases}$$

$$binary \begin{cases} \mathscr{W} ---- \text{ unless} \\ \mathscr{U} ---- \text{ until} \end{cases}$$

together with the fixed point constructors

$$\mu\xi.\chi(\xi) \quad \text{the minimal solution to } \chi(\xi) \Rightarrow \xi$$

$$\nu\xi.\chi(\xi) \quad \text{the maximal solution to } \xi \Rightarrow \chi(\xi).$$

In the fixed-point schema above, ξ is a temporal variable that is bound to some μ or ν quantifier. $\chi(\xi)$ denotes a temporal formula that is free in the temporal variable ξ. Furthermore, for notational convenience, given some fixed point $\mu\xi.\chi(\xi)$, $\chi(\phi)$ denotes the temporal formula $\chi(\xi)$ with all occurrences of ξ replaced by ϕ, i.e. $\chi(\phi/\xi)$.

2.2.2   Models and interpretations

Temporal formulae are interpreted in models M

$$M = \langle \sigma, \alpha, i \rangle \in ((\mathbb{N} \to 2^{\mathscr{A}_L}) \times (\mathbb{N} \to 2^{\mathscr{A}_G}) \times \mathbb{N})$$

σ is a total mapping from $\mathbb{N}$ (the natural numbers) to states (or worlds); each state being a set of local propositions. Such a mapping is often referred to as a state sequence and represented by:

$$s_0 \quad s_1 \quad s_2 \quad \ldots$$

α provides an interpretation for quantified propositions and is similar to σ in being a total mapping over $\mathbb{N}$ to sets of propositions, but propositions here drawn from the alphabet $\mathscr{A}_G$.

i records the index value to be used as *now*, i.e. the present moment in time. Thus if, for proposition p drawn from $\mathscr{A}_L$, $p \in \sigma(i)$ then p is true now. The notation

$$\sigma, \alpha, i \vDash \phi$$

is used to give the interpretation of the temporal formula ϕ in model M; it yields a value **t** or **f**.

The definition of satisfiability, i.e. $\sigma, \alpha, i \vDash \phi$, is now given by induction over the structure of the temporal formula ϕ.

Truthvalues

$$\sigma, \alpha, i \vDash \textbf{true} = \textbf{t} \qquad \sigma, \alpha, i \vDash \textbf{false} = \textbf{f}$$

State propositions $p \in \mathscr{A}_L$

$$\sigma, \alpha, i \vDash p = \begin{cases} \textbf{t} & \text{if } p \in \sigma(i) \\ \textbf{f} & \text{otherwise} \end{cases}$$

Quantified propositions $q \in \mathscr{A}_G$

$$\sigma, \alpha, i \vDash q = \begin{cases} \textbf{t} & \text{if } q \in \alpha(i) \\ \textbf{f} & \text{otherwise} \end{cases}$$

Non-temporal connectives

$$\sigma, \alpha, i \vDash \neg\phi = \begin{cases} \textbf{t} & \text{if } \sigma, \alpha, i \vDash \phi = \textbf{f} \\ \textbf{f} & \text{otherwise} \end{cases}$$

$$\sigma, \alpha, i \vDash \phi \vee \psi = \begin{cases} \textbf{t} & \text{if } \sigma, \alpha, i \vDash \phi = \textbf{t} \text{ or } \sigma, \alpha, i \vDash \psi = \textbf{t} \\ \textbf{f} & \text{otherwise} \end{cases}$$

$$\sigma, \alpha, i \vDash \phi \wedge \psi = \begin{cases} \textbf{t} & \text{if } \sigma, \alpha, i \vDash \phi = \textbf{t} \text{ and } \sigma, \alpha, i \vDash \psi = \textbf{t} \\ \textbf{f} & \text{otherwise} \end{cases}$$

$$\sigma, \alpha, i \vDash \phi \Rightarrow \psi = \sigma, \alpha, i \vDash \neg\phi \vee \psi$$

$$\sigma, \alpha, i \vDash \phi \Leftrightarrow \psi = \sigma, \alpha, i \vDash (\phi \Rightarrow \psi) \wedge (\psi \Rightarrow \phi)$$

$$\sigma, \alpha, i \vDash \exists q . \phi = \begin{cases} \mathbf{t} & \text{if for some } \alpha' \text{ differing from } \alpha \text{ just on } q, \quad \sigma, \alpha', i \vDash \phi = \mathbf{t} \\ \mathbf{f} & \text{otherwise} \end{cases}$$

$$\sigma, \alpha, i \vDash \forall q . \phi = \begin{cases} \mathbf{t} & \text{if for all } \alpha' \text{ differing from } \alpha \text{ just on } q, \quad \sigma, \alpha', i \vDash \phi = \mathbf{t} \\ \mathbf{f} & \text{otherwise} \end{cases}$$

Temporal connectives

$$\sigma, \alpha, i \vDash \bigcirc \phi = \sigma, \alpha, i+1 \vDash \phi$$

$$\sigma, \alpha, i \vDash \Box \phi = \begin{cases} \mathbf{t} & \text{if for every } k \geq i \quad \sigma, \alpha, k \vDash \phi = \mathbf{t} \\ \mathbf{f} & \text{otherwise} \end{cases}$$

$$\sigma, \alpha, i \vDash \Diamond \phi = \begin{cases} \mathbf{t} & \text{if for some } k \geq i \quad \sigma, \alpha, k \vDash \phi = \mathbf{t} \\ \mathbf{f} & \text{otherwise} \end{cases}$$

$$\sigma, \alpha, i \vDash \phi \, \mathcal{U} \, \psi = \begin{cases} \mathbf{t} & \text{if there is some } k \geq i \text{ such that } \sigma, \alpha, k \vDash \psi = \mathbf{t} \\ & \text{and for every } j, \, i \leq j < k, \quad \sigma, \alpha, j \vDash \phi = \mathbf{t} \\ \mathbf{f} & \text{otherwise} \end{cases}$$

$$\sigma, \alpha, i \vDash \phi \, \mathcal{W} \, \psi = \begin{cases} \mathbf{t} & \text{if } \sigma, \alpha, i \vDash \Box \phi = \mathbf{t} \text{ or } \sigma, \alpha, i \vDash \phi \, \mathcal{U} \, \psi = \mathbf{t} \\ \mathbf{f} & \text{otherwise} \end{cases}$$

Fixed-point constructors

The formation of $\chi(\xi)$ is restricted so that occurrences of ξ in $\chi(\xi)$ appear under an even number of negations. This is a monotonicity restriction which ensures the existence of minimal and maximal solutions to $\chi(\xi) \Rightarrow \xi$ and $\xi \Rightarrow \chi(\xi)$ respectively. Furthermore, the minimal and maximal fixed-point solutions can be characterized in terms of limits of successive approximations.

$$\sigma, \alpha, i \vDash \nu \xi . \chi(\xi) = \begin{cases} \mathbf{t} & \text{if for some limit ordinal } \lambda \quad \sigma, \alpha, i \vDash \chi_\wedge^\lambda(\mathbf{true}) = \mathbf{t} \\ \mathbf{f} & \text{otherwise} \end{cases}$$

$$\sigma, \alpha, i \vDash \chi_\wedge^\lambda(\xi) = \begin{cases} \mathbf{t} & \text{if for every ordinal } o < \lambda \quad \sigma, \alpha, i \vDash \chi_\wedge^o(\xi) = \mathbf{t} \\ \mathbf{f} & \text{otherwise} \end{cases}$$

$$\sigma, \alpha, i \vDash \mu \xi . \chi(\xi) = \begin{cases} \mathbf{t} & \text{if for some limit ordinal } \lambda \quad \sigma, \alpha, i \vDash \chi_\vee^\lambda(\text{false}) = \mathbf{t} \\ \mathbf{f} & \text{otherwise} \end{cases}$$

$$\sigma, \alpha, i \vDash \chi_\vee^\lambda(\xi) = \begin{cases} \mathbf{t} & \text{if for some ordinal } o < \lambda \quad \sigma, \alpha, i \vDash \chi_\vee^o(\xi) = \mathbf{t} \\ \mathbf{f} & \text{otherwise} \end{cases}$$

For non-limit ordinals n,

$$\chi_\wedge^{n+1}(\xi) \stackrel{\text{def}}{=} \chi(\chi_\wedge^n(\xi)) \text{ and similarly } \chi_\vee^{n+1}(\xi) \stackrel{\text{def}}{=} \chi(\chi_\vee^n(\xi))$$

where

$$\chi_\wedge^0(\xi) = \chi_\vee^0(\xi) = \xi.$$

Fixpoint theory has shown that under certain continuity restrictions iteration is only necessary to the first limit ordinal ω. Here, two types of continuity are important. A function χ is said to be $\wedge$-continuous if

$$\chi\left(\bigwedge_{n<\omega} \varphi_n\right) \equiv \bigwedge_{n<\omega} \chi(\varphi_n)$$

So, if χ is monotonic and $\wedge$-continuous then

$$\sigma, \alpha, i \vDash \nu\xi \, . \, \chi(\xi) = \begin{cases} \mathbf{t} & \text{if for all } n \; \sigma, \alpha, i \vDash \chi_\wedge^n(\mathbf{true}) = \mathbf{t} \\ \mathbf{f} & \text{otherwise} \end{cases}.$$

Similarly for the definition of minimal fixed points.

A temporal formula ϕ is said to be *true* for model $M = \langle \sigma, \alpha, i \rangle$ (alternatively, *satisfied* by M) if and only if $\sigma, \alpha, i \vDash \phi = \mathbf{t}$.

A temporal formula ϕ is then defined to be *valid*, denoted by $\vDash \phi$, if and only if ϕ is true for every model M.

2.2.3 Axioms and proof rules

An outline of an axiomatization for this logic is presented, for convenience, in terms of $\bigcirc$, ν and $\forall$.[5] The given axiomatization does not cater for solutions to sets of simultaneous recursive definitions; such problems are discussed in Banieqbal and Barringer (1986) where issues of completeness are considered.

Axioms

$$\vdash w \quad \text{propositional tautology}$$

$$\vdash \neg \bigcirc \phi \Leftrightarrow \bigcirc \neg \phi$$

$$\vdash \bigcirc(\phi \Rightarrow \psi) \Rightarrow (\bigcirc \phi \Rightarrow \bigcirc \psi)$$

$$\vdash \nu\xi \, . \, \chi(\xi) \Rightarrow \chi(\nu\xi \, . \, \chi(\xi))$$

$$\vdash \forall q \, . \, \phi \Rightarrow \phi[p/q]^6$$

$$\vdash \forall q \, . \, \nu\xi \, . \, \chi(\xi) \Rightarrow \nu\xi \, . \, \forall q \, . \, \chi(\xi)$$

---

5. Strictly, an axiomatization could be presented in terms of $\bigcirc$ and ν.
6. This axiomatization further assumes that substitution is a meta-level notion and not handled via this proof system.

Inference rules

Modus Ponens (MP)

$$\frac{\vdash \phi \qquad \vdash \phi \Rightarrow \psi}{\vdash \psi}$$

Next Introduction ($\bigcirc$-I)

$$\frac{\vdash \phi}{\vdash \bigcirc \phi}$$

ν Introduction (ν-I)

$$\frac{\vdash \theta \Rightarrow \chi(\theta)}{\vdash \theta \Rightarrow \nu \xi . \chi(\xi)}$$

Universal Introduction ($\forall$-Intro)

$$\frac{\vdash \phi \Rightarrow \psi}{\vdash \phi \Rightarrow \forall q . \psi[q/p]}$$

where p does not occur free in ϕ and q occurs free for p in ψ and is drawn from $\mathcal{A}_G$.

Definitions

$$\mu \xi . \chi(\xi) \stackrel{\text{def}}{=} \neg \nu \xi . \neg \chi(\neg \xi)$$

$$\phi \mathcal{W} \psi \stackrel{\text{def}}{=} \nu \xi . (\psi \vee \phi \wedge \bigcirc \xi)$$

$$\phi \mathcal{U} \psi \stackrel{\text{def}}{=} \mu \xi . (\psi \vee \phi \wedge \bigcirc \xi)$$

$$\square \phi \stackrel{\text{def}}{=} \phi \mathcal{W} \textbf{false}$$

$$\Diamond \phi \stackrel{\text{def}}{=} \neg \square \neg \phi$$

$$\exists q . \phi \stackrel{\text{def}}{=} \neg \forall q . \neg \phi$$

2.2.4 Comments

In the above definitions, note the relation between maximal and minimal fixed points. Furthermore, notice how $\phi \mathcal{W} \psi$ is the maximal solution to the equivalence $\xi \Leftrightarrow \psi \vee \phi \wedge \bigcirc \xi$ and that $\phi \mathcal{U} \psi$ is its minimal solution.

Below, four useful theorems are noted.

$$\vdash \Box \phi \Leftrightarrow \nu \xi . \phi \wedge \bigcirc \xi$$

$$\vdash \Diamond \phi \Leftrightarrow \mu \xi . \phi \vee \bigcirc \xi$$

$$\vdash \nu \xi . \chi(\xi) \Leftrightarrow \chi(\nu \xi . \chi(\xi))$$

$$\vdash \mu \xi . \chi(\xi) \Leftrightarrow \chi(\mu \xi . \chi(\xi))$$

These theorems, for which proofs are left as an exercise to the interested, are of use in later examples.

2.3 Action language semantics

2.3.1 General approach

A temporal formula identifies a set of models, those models that satisfy the formula. As indicated above, a temporal logic model here includes a 'sequence' of sets of local propositions, the σ component. It is this component that enables certain of these models to be viewed as execution traces, or behaviours, of an action language statement (or program). The execution of some action, $a \in \mathcal{A}$, is identified with a unique local proposition, say $a \in \mathcal{A}_L$, being true; the proposition is false if the action is not happening. Thus, given some behaviour, i.e. a sequence of action names, a corresponding temporal logic model can be constructed, i.e. a sequence of sets of propositions with each set identifying the actions that occurred at that 'time'. Whether a temporal model corresponds to some behaviour of some action language statement depends, naturally, upon the details of the computation model. In a strictly interleaved model of parallel computation, for example, only sequences of sets of cardinality one will correspond to behaviours as only one action can occur at any one time.

A *temporal* semantics for the action language is given by providing some function $\mathcal{F}$ which, essentially, maps an action language statement, or program, into a temporal formula, the models of which 'correspond' to the behaviour of that object. It is, of course, desirable that such a description is compositional, within the temporal framework. This means that the temporal formula denoting the semantics of some composite language fragment must be given as some temporal composition of the temporal formulae denoting the semantics of the immediate syntactic subcomponents of the composite fragment. For example, a temporal formula denoting the semantics of $S_1 \parallel S_2$ must be given by some logical composition of the temporal formulae denoting S_1's and S_2's semantics.

Consider now the semantics of parallel composition in the action language in more detail. The example of Section 2.1 indicated that the parallel composition of S_1 and S_2 should be taken as all possible fair interleavings of the action traces of S_1 with those of S_2. If the temporal semantics of some construct S just describes traces of actions of S, the temporal language would clearly require an operator corresponding to fair interleaving of models in order to be compositional. Such an operator is not available! For this reason a different approach is adopted.

The notion of *open* semantic descriptions is used. An open temporal semantics of some construct S describes the action traces of S in all possible (parallel) environments. To achieve this, a special proposition i, representing an unknown environment action, is introduced. For example, if the action trace of S is a^ω, the open temporal semantics of S will describe the traces $(i^*a)^\omega$. Of course, at any stage the semantics of the construct can be *closed* by simply removing traces with i actions appearing.[7] Furthermore, an open semantics of some construct can be partially closed when part of its environment becomes known. For example, the open semantics of $S_1 \parallel S_2$ is obtained by taking the conjunction of the partially closed semantics of S_1 and S_2. That is to say, the open semantics of S_1 and of S_2 are modified to reflect that S_2 is part of S_1's environment, and vice versa, by changing i in the S_1 semantics to be either i or some action of S_2, and vice versa. This technique thus enables a compositional temporal semantics to be given.

2.3.2 Action language temporal semantics

Consider now, formally, a compositional temporal semantics for the action language introduced in Section 2.1. The temporal logic introduced in Section 2.2 must be *tailored* so that there is a straightforward correspondence between temporal models and action language computation models.

The alphabet of local propositions, $\mathscr{A}_L$, is taken as the union of the action alphabet, $\mathscr{A}$, and the test alphabet, $\mathscr{T}$, together with the unique proposition i, i.e. $i \notin \mathscr{A} \cup \mathscr{T}$.

To enforce a strict interleaving model of parallel computation, it is necessary to assume that at most one proposition from the $\mathscr{A}$ subset of $\mathscr{A}_L$ is

---

7. In terms of temporal logic, if $\mathscr{F}(S)$ denotes the open semantics, then $\mathscr{F}(S) \wedge \square \neg i$ will denote the closed semantics of S.

true at any moment in time. The temporal logic is thus modified by the following constraint on propositions,

$$\vDash \left(\sum_{a \in \mathcal{A}} a \right) \leqslant 1$$

where **true, false** are treated as 1 and 0 respectively.[8] There also are likely to be relations (temporal) between actions and tests; these are tactitly assumed.

Statements

The temporal semantic function $\mathcal{F}$, giving open semantic denotations to statements, has signature

$$\mathcal{F} : \mathcal{S} \rightarrow \mathcal{E} \rightarrow \mathcal{T}\mathcal{L}.^{9}$$

As action language statements execute in a procedural context, providing definitions of procedures, it is necessary to parametrize the meaning function with such an object containing the semantics of procedures. In the signature of $\mathcal{F}$ above, $\mathcal{E}$ is the space of mappings from procedure names to temporal formulae, i.e. $\mathcal{E} = \mathcal{P} \rightarrow \mathcal{T}\mathcal{L}$. Throughout the following it is assumed that Θ is a typical mapping from the space $\mathcal{E}$. $\mathcal{F}(S)$ is, therefore, a function mapping procedural environments to temporal formulae, and, hence, when supplied with some Θ, i.e. $(\mathcal{F}(S))(\Theta)$, yields a temporal formula denoting the temporal semantics of statement S in the procedural environment Θ. For notational convenience below, $\mathcal{F}_\Theta(S)$ is used instead of $(\mathcal{F}(S))(\Theta)$ and, for $p \in \mathcal{P}$, Θ_p is used instead of $\Theta(p)$. The 'semantic equations' for statements are listed below.

$$\mathcal{F}_\Theta(p) \qquad\qquad \overset{\text{def}}{=} \Theta_p$$

$$\mathcal{F}_\Theta(a.S) \qquad\qquad \overset{\text{def}}{=} i\mathcal{U}(a \wedge \bigcirc\mathcal{F}_\Theta(S))$$

$$\mathcal{F}_\Theta(\square_{k \in \kappa}\, t_k \rightarrow S_k) \overset{\text{def}}{=} \underbrace{\left(i\mathcal{U}\left(\bigvee_{k \in \kappa} (t_k \wedge \mathcal{F}_\Theta(S_k)) \right) \right)}_{\text{success}} \vee \underbrace{\square\left(i \wedge \Diamond\left(\bigwedge_{k \in \kappa} \neg t_k \right) \right)}_{\text{failure}}$$

$$\mathcal{F}_\Theta(S_1 \| S_2) \qquad \overset{\text{def}}{=} \mathcal{F}_\Theta(S_1)[i \vee \bar{\alpha}(S_2)/i] \wedge \mathcal{F}_\Theta(S_2)[i \vee \bar{\alpha}(S_1)/i]$$

---

8. Of course, the language used to express this constraint is not part of the language described in Section 2.2.1. It is felt, however, that this form is rather easier to understand than stricter realizations using logical operators.

9. $\rightarrow$ associates to the right; thus $\mathcal{F} : \mathcal{S} \rightarrow (\mathcal{E} \rightarrow \mathcal{T}\mathcal{L})$.

where

$$\bar{\alpha}(S) \stackrel{\text{def}}{=} \bigvee_{a \in \alpha(S)} a$$

To help the reader comprehend this semantic description, an informal description of the equations follows.

The first equation is straightforward. The value of Θ at p, i.e. Θ_p, is by definition the open temporal semantics of procedure p!

The open semantics for $a . S$, atomic sequential composition, allows for an arbitrary finite number of (arbitrary) environment actions to occur before the action a. The action a is immediately followed by the open semantics of S, given by $\mathcal{F}_\Theta(S)$. Note that the semantics makes use of $\mathcal{U}$ and thus guarantees that a will eventually occur. This, therefore, implies a fair interleaving of the parallel environment with the construct $a . S$.

The guarded choice open semantics is a little harder! Execution of the guarded choice statement may be *successful* in that some guard t_k evaluates to true thus enabling the branch statement S_k to be executed, or it may lead to a *blocked situation* where all guards evaluate to false. This is reflected in the semantic description by two clauses, each, respectively, reflecting the above behaviours. The first clause of the semantics, covering successful passing, allows arbitrary environment actions until some guard, say t_k, eventually evaluates to true, in which case the open semantics of the corresponding statement S_k follows. A blocked situation, i.e. failure to pass any guard, covered by the second clause, is described by there being no action of the choice statement but at least infinitely many moments when all guards evaluate to false, these being moments when the guards happened to be (virtually) tested.

To some extent such a description of an open semantics for guarded choice is unsatisfactory. The open semantics of a blocked choice will be

$$\Box\left(i \wedge \Diamond\left(\bigwedge_{k \in \kappa} \neg t_k\right)\right)$$

and therefore its closed semantics, obtained by eliminating environment actions, is

$$\Box\left(i \wedge \Diamond\left(\bigwedge_{k \in \kappa} \neg t_k\right)\right) \wedge \Box \neg i$$

which is clearly **false**. This problem stems from the decision to make testing of guards appear to take no time, which is, of course, a little unrealistic. In Section 2.5 a better solution is presented.

The open semantics for parallel composition is given as described informally above. An example at the end of this section on semantics will help to justify the approach.

Programs and procedures

The temporal semantic function $\mathcal{M}$,

$$\mathcal{M} : \mathcal{PR} \to \mathcal{TL}$$

provides (temporal) meaning to programs $Pr \in \mathcal{PR}$.

For simplicity, consider the open temporal semantics of a program containing just one procedure definition.

$$\mathcal{M}\left(p \overset{\text{def}}{=} S_p; S \right) \overset{\text{def}}{=} \mathcal{F}_\Theta(S)$$

where the procedural environment Θ is defined as the mapping

$$[p \to \nu\xi . \mathcal{F}_{[p \to \xi]}(S_p)].$$

This defines that the semantics of p is the maximal solution to the temporal equivalence

$$\xi \Leftrightarrow \mathcal{F}_{[p \to \xi]}(S_p).$$

Assuming ξ is the open semantics of a call of p, then this semantics should be equivalent to the open semantics of the body of p, S_p, evaluated in the procedural environment that maps p to the open semantics ξ.

The temporal semantics for programs with more than one procedure definition is obtained from the obvious extension of the above.

2.3.3 Example

Consider the trivial program introduced in Section 2.1.

$$p \overset{\text{def}}{=} a.p; \quad q \overset{\text{def}}{=} b.q; \quad p \| q$$

The open semantics for procedure p, i.e. $\nu\xi . \mathcal{F}_{[p \to \xi]}(a.p)$, is given as $\nu\xi . i\mathcal{U}(a \wedge \bigcirc \xi)$. Similarly, the open semantics for q is given as $\nu\xi . i\mathcal{U}(b \wedge \bigcirc \xi)$.

The open semantics of $p \| q$ should therefore be evaluated in the procedural context of Θ where

$$\Theta \overset{\text{def}}{=} [p \to \nu\xi . i\mathcal{U}(a \wedge \bigcirc \xi), q \to \nu\xi . i\mathcal{U}(b \wedge \bigcirc \xi)]$$

and hence, following the above descriptions, given by

$$(\nu\xi . i\mathcal{U}(a \wedge \bigcirc \xi))[i \vee \bar{\alpha}(q)/i] \wedge (\nu\xi . i\mathcal{U}(b \wedge \bigcirc \xi))[i \vee \bar{\alpha}(p)/i]$$

Clearly, $\alpha(p)$, denoting the minimal solution to $\alpha(p) = \{\alpha\} \cup \alpha(p)$, is $\{a\}$ and hence $\bar{\alpha}(p)$ is just the proposition a. Similarly $\bar{\alpha}(q)$ is b. Rewriting the above yields

$$(\nu\xi . i\mathcal{U}(a \wedge \bigcirc\xi))[i \vee b/i] \wedge (\nu\xi . i\mathcal{U}(b \wedge \bigcirc\xi))[i \vee a/i]$$

and therefore, as substitution distributes across ν (μ),

$$\mathcal{F}_{\Theta}(p \| q) \overset{\text{def}}{=} \underbrace{\nu\xi . (i \vee b)\mathcal{U}(a \wedge \bigcirc\xi)}_{F_{p'}} \wedge \underbrace{\nu\xi . (i \vee a)\mathcal{U}(b \wedge \bigcirc\xi)}_{F_{q'}}$$

1. It is a simple matter to show that $F_{p'} \wedge F_{q'} \Rightarrow \nu\xi . i\mathcal{U}((a \vee b) \wedge \bigcirc\xi)$, for

$$F_{p'} \wedge F_{q'} \Leftrightarrow (i \vee b)\mathcal{U}(a \wedge \bigcirc F_{p'}) \wedge (i \vee a)\mathcal{U}(b \wedge \bigcirc F_{q'}) \tag{1}$$

$$\Leftrightarrow i\mathcal{U}(a \wedge \bigcirc(F_{p'} \wedge (i \vee a)\mathcal{U}(b \wedge \bigcirc F_{q'})))$$
$$\vee$$
$$i\mathcal{U}(b \wedge \bigcirc(F_{q'} \wedge (i \vee b)\mathcal{U}(a \wedge \bigcirc F_{p'}))) \tag{2}$$

$$\Leftrightarrow i\mathcal{U}((a \vee b) \wedge \bigcirc(F_{p'} \wedge F_{q'})) \tag{3}$$

$$F_{p'} \wedge F_{q'} \Rightarrow \boxed{\nu\xi . i\mathcal{U}((a \vee b) \wedge \bigcirc\xi)} \tag{4}$$

Line (1) is justified by the fixed-point theorem, $F_{p'}$ and $F_{q'}$ having been unravelled once. Line (2) is justified by the semantic requirement that the propositions a and b are never true together. Line (3) is then a refolding, i.e. substitution of the fixed points. Finally, line (4) is justified by the fixed-point introduction. $F_{p'} \wedge F_{q'}$ is a solution of $\xi \Rightarrow i\mathcal{U}((a \vee b) \wedge \bigcirc\xi)$ and must therefore be not greater than the maximal solution.

2. From line (4) above, it is straightforward to show that the closed temporal semantics for the example program implies $\square(a \vee b)$. It is necessary to show that

$$\underbrace{\nu\xi . i\mathcal{U}((a \vee b) \wedge \bigcirc\xi)}_{P} \wedge \underbrace{\nu\xi . \neg i \wedge \bigcirc\xi}_{Q} \Rightarrow \square(a \vee b)$$

Well, letting P and Q denote the fixed points as indicated above, P and Q can be unfolded using the fixed point theorem to yield

$$P \wedge Q \Leftrightarrow i\mathcal{U}((a \vee b) \wedge \bigcirc P) \wedge \neg i \wedge \bigcirc Q$$

which, after unfolding $\mathcal{U}$ once, leads to

$$P \wedge Q \Leftrightarrow ((a \vee b) \wedge \bigcirc P \vee i \wedge \bigcirc(i\mathcal{U}((a \vee b) \wedge \bigcirc P))) \wedge \neg i \wedge \bigcirc Q.$$

Therefore after distributing $\wedge$ over $\vee$,

$$P \wedge Q \Leftrightarrow \neg i \wedge (a \vee b) \wedge \bigcirc(P \wedge Q)$$

which after weakening yields

$$P \wedge Q \Rightarrow (a \vee b) \wedge \bigcirc(P \wedge Q)$$

and hence, by maximal fixed point introduction,

$$P \wedge Q \Rightarrow \nu\xi.(a \vee b) \wedge \bigcirc\xi$$

which is, of course, by definition of $\square$, the desired result.

$$\boxed{P \wedge Q \Rightarrow \square(a \vee b)}$$

3. A stronger specification of the example program is to show that both a and b must occur infinitely often, i.e. $\square(\lozenge a \wedge \lozenge b)$. First note that $\lozenge F_{p'} \Rightarrow \lozenge a \wedge \bigcirc\lozenge F_{p'}$, for

$$\lozenge F_{p'} \Leftrightarrow \lozenge((i \vee b)\mathcal{U}(a \wedge \bigcirc F_{p'}))$$
$$\Rightarrow \lozenge(\lozenge(a \wedge \bigcirc F_{p'}))$$
$$\Rightarrow \lozenge(a \wedge \bigcirc F_{p'})$$
$$\Rightarrow \lozenge a \wedge \lozenge\bigcirc F_{p'}$$
$$\Rightarrow \lozenge a \wedge \bigcirc\lozenge F_{p'}$$

Similarly

$$\lozenge F_{q'} \Rightarrow \lozenge b \wedge \bigcirc\lozenge F_{q'}$$

Therefore,

$$\lozenge F_{p'} \wedge \lozenge F_{q'} \Rightarrow (\lozenge a \wedge \lozenge b) \wedge \bigcirc(\lozenge F_{p'} \wedge F_{q'})$$

and so, by fixed point introduction,

$$\lozenge F_{p'} \wedge \lozenge F_{q'} \Rightarrow \nu\xi.(\lozenge a \wedge \lozenge b) \wedge \bigcirc\xi$$

i.e., by definition of $\square$ as fixed point,

$$\lozenge F_{p'} \wedge \lozenge F_{q'} \Rightarrow \square(\lozenge a \wedge \lozenge b)$$

As $\phi \wedge \psi \Rightarrow \lozenge\phi \wedge \lozenge\psi$ the desired result follows by Modus Ponens.

$$\boxed{F_{p'} \wedge F_{q'} \Rightarrow \square(\lozenge a \wedge \lozenge b)}$$

2.4 Action Language Proof System

Having produced a compositional temporal semantic description of the action language, a compositional temporal proof system for that language can be obtained most easily. The system is presented in the form of a programming logic—in particular, the notation is based on that of 'Hoare Logic', Hoare (1969).

The programming logic is an extension of the $\mathcal{TL}$ logic introduced in Section 2.2. The system is extended with sentences 'denoting', typically, that a temporal formula ϕ is a specification of some language statement S. The following notation is used

$$\vdash \{S\}_\Theta \; \varphi$$

to indicate that a proof exists to show that φ is a specification of the statement S assuming procedural context Θ, i.e. all open executions of S are described by φ. Indeed, the connection with the above temporal semantics is simply

$$\text{If } \vdash \{S\}_\Theta \; \varphi \text{ then } \vdash \mathscr{F}_\Theta(S) \Rightarrow \varphi.[10]$$

The following axioms and rules form the system.

Interleaving

$$\vdash \left(\sum_{a \in \mathscr{A}} a \right) \leqslant 1.[11]$$

This corresponds to the model requirement given earlier that only one local 'action' proposition is true at any one time.

Consequence

$$\frac{\vdash \{S\}_\Theta \; \varphi \quad \vdash \varphi \Rightarrow \psi}{\vdash \{S\}_\Theta \; \psi}$$

A standard weakening rule.

Atomic composition

$$\frac{\vdash \{S\}_\Theta \; \varphi}{\vdash \{a \, . \, S\}_\Theta \; i \mathscr{U}(a \wedge \bigcirc \varphi)}$$

If φ is a specification of S then clearly prefixing φ by $i \mathscr{U} a$ will be a valid open specification of $a \, . \, S$.

Guarded Choice

$$\frac{\vdash \{S_k\}_\Theta \; \varphi_k \quad k \in K}{\vdash \{\Box_{k \in K} \, t_k \to S_k\}_\Theta (i \mathscr{U}(\bigvee_{k \in K}(t_k \wedge \varphi_k))) \vee \Box(i \wedge \Diamond(\bigwedge_{k \in K} \neg t_k))}$$

This rule follows from replacing the semantics for the statements S_i by their respective specifications ϕ_i in the semantics for guarded choice.

---

10. Actually, $\vdash \mathscr{F}_\Theta(S) \Rightarrow \phi$ could equally be used given an complete axiomatization of $\mathscr{T}\mathscr{L}$.
11. Again, this notation is used for simplicity and readability.

Parallel composition

$$\frac{\vdash\{S_j\}_\Theta\; \varphi_j \quad j = 1, 2}{\vdash\{S_1 \| S_2\}_\Theta\; \varphi_1[i \lor \bar{\alpha}(S_2)/i] \land \varphi_2[i \lor \bar{\alpha}(S_1)/i]}$$

Again, this rule follows directly from the semantic description for parallel composition given above.

Program (simplified)

$$\frac{\vdash\{S_p\}_{[p\to\xi]}\varphi(\xi) \quad \vdash\{S\}_{[p\to v\xi.\varphi(\xi)]}\psi}{\vdash\left\{p \stackrel{\text{def}}{=} S_p; S\right\}\psi}$$

This rule assumes only one procedure definition as the context of the statement S. If $\varphi(\xi)$ can be established for the body of the procedure p under the assumption that ξ is the meaning of the procedure, then $v\xi.\varphi(\xi)$ will certainly be a specification of a call of the procedure. So if, under that assumption, ψ can be established as a specification of the statement S, then it will be a specification for the given program. It is relatively straightforward to extend this to cater for more than one procedure definition.

2.5 Introducing Hiding

A feature of most programming languages is a mechanism to control the *scope* of program variables, i.e. from where a particular variable can be accessed. Outside the scope of a variable, that variable is hidden. In the action language, where variables do not exist as such, the notion of scoping can be applied directly to actions. The action language is extended with a restriction operation \. The alphabet of $S\backslash a$, $\alpha(S\backslash a)$, is $\alpha(S) - \{a\}$, and, informally, the semantics of the construct $S\backslash a$ is the semantics of S with all a actions hidden (or internalized).

It is natural to imagine that the open temporal semantics of $S\backslash a$ could be taken as $\exists a.\mathcal{F}_\Theta(S)$. However, adopting such a semantics introduces the possibility of non-interleaved actions. The mixture of non-interleaving (for hidden actions) and interleaving (for visible actions) semantics leads to unfortunate non-identities. For example, if a is an action of S_1 (and therefore not of S_2) and b is an action of S_2, then it should follow that

$$\mathcal{F}_\Theta(S_1\backslash a \| S_2\backslash b) \Leftrightarrow \mathcal{F}_\Theta((S_1 \| S_2)\backslash a\backslash b)$$

However, this is not the case.

It is therefore necessary to introduce the notion of an *internal* unnamed action, similar to the τ action of CCS, see Milner (1980); in fact the same name is used and the proposition τ is taken as a unique local proposition different from i and not in $\mathcal{A} \cup \mathcal{T}$. The semantics of hiding is then taken as:

$$\mathcal{F}_\Theta(S \backslash a) \stackrel{\text{def}}{=} \mathcal{F}_\Theta(S)[\tau/a]$$

i.e. the semantics of S with all a actions replaced by τ actions.

Such a change requires that the semantics of parallel composition also change.

$$\mathcal{F}_\Theta(S_1 \| S_2) \stackrel{\text{def}}{=} \exists \tau_1, \tau_2 . \{\Box((\tau \Leftrightarrow \tau_1 \vee \tau_2) \wedge \neg(\tau_1 \wedge \tau_2)) \wedge$$

$$\mathcal{F}_\Theta(S_1)[\tau_1/\tau][i \vee \tau_2 \vee \bar{\alpha}(S_2)/i] \wedge$$

$$\mathcal{F}_\Theta(S_2)[\tau_2/\tau][i \vee \tau_1 \vee \bar{\alpha}(S_1)/i]\}$$

The introduction of the hidden internal action now enables a better description for the blocked behaviour of a guarded choice. The earlier semantics assumed testing could be achieved at no cost. The obvious remedy is to make a test of the guards into a τ action. For example,

$$\mathcal{F}_\Theta(\Box_{k \in \kappa} t_k \rightarrow S_k) \stackrel{\text{def}}{=} \underbrace{\left(i\mathcal{U}\left(\tau \wedge \bigwedge_{k \in \kappa} \neg t_k\right)\right)}_{\text{failure}} \mathcal{W} \underbrace{\left(\bigvee_{k \in \kappa}(t_k \wedge \tau \wedge \bigcirc \mathcal{F}_\Theta(S_k))\right)}_{\text{success}}$$

The successful part starts with a silent τ action and some test t_k evaluating true and is then followed by the semantics of the appropriate branch S_k. A blocked guarded choice is described by there being infinitely many τ actions with *all* the guards false; other moments are i actions.

3   FURTHER DEVELOPMENTS

In the previous section an *action language* is used as a vehicle to demonstrate basic techniques for obtaining compositional temporal semantic descriptions and language proof theories. Figure 3 highlights various developments that can be made from the action language base. The left side of the figure considers developments towards full shared-variable programming languages, that is, programming languages where communication between parallel processes is achieved through global, shared, variables. The right side provides development towards an alternative, more structured, form of communication mechanism, i.e. synchronous message passing. Naturally, some languages, such as Ada, do bridge the branches!

Some of the developments are considered briefly in this section.

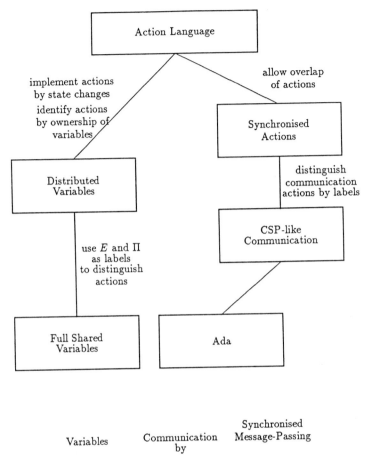

Fig. 3. Language developments.

3.1 Distributed Variables

The language handled here can describe processes that communicate through shared variables, but with the strong restriction that sharing is limited to *reading* other processes' variables (a process may only write to the variables it has declared it *owns*); such a restriction eliminates most of the difficulties of *interference*. The programming language, otherwise, has the usual constructs, namely, assignment, guarded choice, parallel composition and recursion.

Again, the temporal language is used to describe the observable behaviour of processes; an observation of a process is, here, simply taken as a snapshot of the state of the variables of that process. The previous temporal language must therefore be modified in the obvious way to reflect the notion of variables, and quantification over values of variables; this is assumed.

In order to describe the behaviour of some program Pr it is necessary to know the sets of variables that can be modified by any particular subcomponent π of Pr. Such variable sets are easily constructed from π and its context; it is assumed, below, that $owns(\pi)$ yields the variable sets of π.[12]

In the previous open semantic descriptions for the action language, the proposition i was used to denote an unknown environment action. In the descriptions given here the notion of unknown environment action is continued but in the guise of an *identity step* of the component process. When a process π is inactive, none of the variables in $owns(\pi)$ can be modified, since π is the only process that can modify them. In this way, an identity step of a process π can be viewed as an unknown environment action of π; the identity step of π places no constraints on the variables owned by π's environment. Assuming $\bar{w}$ ($\bar{u}$) denotes a set of variables (values), the predicate $I(\bar{w})$ is defined as an identity step with respect to the variables $\bar{w}$, i.e.

$$I(\bar{w}) \stackrel{\text{def}}{=} \exists \bar{u} . \bar{w} = \bar{u} \wedge \bigcirc(\bar{w} = \bar{u}).$$

Furthermore, for ease of notation, I is used to denote $I(owns(\pi))$. In the semantic equations below, I replaces the previous use of the environmental proposition i.

$$\mathscr{F}_\Theta(p) \stackrel{\text{def}}{=} \Theta_p$$

$$\mathscr{F}_\Theta(y := e; S) \stackrel{\text{def}}{=} I\mathcal{U}(\bigcirc(y = e) \wedge I(owns(\pi) - \{y\}) \wedge \bigcirc\mathscr{F}_\Theta(S))$$

$$\mathscr{F}_\Theta(\square_{k \in \kappa} t_k \to S_k) \stackrel{\text{def}}{=} I\mathcal{U}((I\mathcal{U}(I \wedge \bigwedge_{k \in \kappa} \neg t_k))\mathcal{W}(I\mathcal{U} \bigvee_{k \in \kappa} (t_k \wedge \mathscr{F}_\Theta(S_k))))$$

$$\mathscr{F}_\Theta(S_1 \| S_2) \stackrel{\text{def}}{=} \mathscr{F}_\Theta(S_1) \wedge \mathscr{F}_\Theta(S_2)$$

Because the variables owned by a process are not modified by any other process, the open description of the assignment statement allows for an arbitrary number of other process steps to precede the updating of the owned variable y.

The semantic descriptions of procedure call and guarded choice are very similar to before.

---

12. Full treatment of this is given in Barringer *et al.* (1986).

However, contrasting with the previous semantics, the parallel composition semantics is much simpler. There is no need, here, to modify explicitly the environment action i in order to let some other process' action occur.

3.2 Fully-shared Variables

To simplify the compositional parallel semantics of the action language, the alphabets of distinct processes were restricted to be disjoint. In the shared variable paradigm, atomic actions are realised as atomic assignments. The distributed variables language of Section 3.1 kept atomic assignments distinct by syntactically ensuring that no two distinct processes could assign to the same variable. In a language with fully-shared variables that admits interference, e.g. two distinct processes writing to the same variables, some additional means of distinguishing atomic actions, i.e. assignments, is required. This extra distinction takes the form of labelling every action in an open computation model of process S as either a Π action, i.e. an action belonging to the process S, or an E action,[13] i.e. an action belonging to one of the processes in S's parallel environment. In the papers of Barringer et al. (1984, 1985) these propositions are taken as transition propositions (propositions that are evaluated on a transition between states, rather than in a state) and the temporal logic is suitably augmented. Assuming such a logic, and a simple shared-variable language with global variables $\bar{y}$, the following equations indicate the form of semantics that is obtained.

$$\mathscr{F}_\Theta(p) \quad \overset{\text{def}}{=} E\mathscr{U}\Theta_p$$

$$\mathscr{F}_\Theta(x := e; S) \quad \overset{\text{def}}{=} E\mathscr{U}\exists\bar{u}.(\Pi \wedge \bar{y} = \bar{u} \wedge \bigcirc(\bar{y} = \bar{u}[e/x] \wedge \mathscr{F}_\Theta(S)))$$

$$\mathscr{F}_\Theta(\square_{k\in\kappa}\, t_k \to S_k) \overset{\text{def}}{=} E\mathscr{U}((E\mathscr{U} \bigwedge_{k\in\kappa} test(\neg t_k))\,\mathscr{W}$$

$$(E\mathscr{U} \bigvee_{k\in\kappa}(test(t_k) \wedge \bigcirc\mathscr{F}_\Theta(S_k))))$$

$$\mathscr{F}_\Theta(S_1 \| S_2) \quad \overset{\text{def}}{=} \exists\Pi_1, \Pi_2.\{\square((\Pi \Leftrightarrow \Pi_1 \vee \Pi_2) \wedge \neg(\Pi_1 \wedge \Pi_2)) \wedge$$

$$\mathscr{F}_\Theta(S_1)[\Pi_1/\Pi] \wedge \mathscr{F}_\Theta(S_2)[\Pi_2/\Pi]\}$$

where

$$test(t) \overset{\text{def}}{=} \Pi \wedge t \wedge \exists\bar{u}.(\bar{y} = \bar{u} \wedge \bigcirc(\bar{y} = \bar{u}))$$

---

13. Actually, E is the complement of Π, and so only the notion of Π proposition is required. For ease, however, E is used instead of $\neg\Pi$.

There is clearly a strong resemblance to the semantic descriptions in the preceding sections. In particular, notice how similar the semantics of the parallel construct is to the parallel semantics given in Section 2.5.

3.3 Synchronized Actions

Consider now moving down the right side of Fig. 3. An obvious first extension to the action language is to allow synchronization of actions from different parallel processes. For example, modify the parallel composition operator to be $\|_\Gamma$ where Γ is an alphabet of actions that must be synchronized. Furthermore, modify the alphabet restriction on $S_1 \|_\Gamma S_2$ to be $\alpha(S_1) \cap \alpha(S_2) \subseteq \Gamma$. The description of parallel composition then becomes

$$\mathscr{F}_\Theta(S_1 \|_\Gamma S_2) \stackrel{\text{def}}{=} \mathscr{F}_\Theta(S_1)[i \vee \bar{\alpha}_\Gamma(S_2)/i] \wedge \mathscr{F}_\Theta(S_2)[i \vee \bar{\alpha}_\Gamma(S_1)/i]$$

where

$$\bar{\alpha}_\Gamma(S) \stackrel{\text{def}}{=} \bigvee_{a \in \alpha(S) - \Gamma} a$$

and is quite self explanatory.

A more complex rule results when hiding is taken in to consideration; this is due to the introduction of the internal action τ.

3.4 Channel-based Communication

In Barringer *et al.* (1985) a compositional temporal semantics for a general CSP-like language is presented. Communication between processes is over channels and process creation is dynamic. To handle channels in the semantics, the notion of transition variable, rather than proposition, is introduced. However, because of the generality of the language and its features, the temporal semantics appears unduly complex, especially the parallel semantics. The report Barringer and Fisher (1987) presents simpler compositional temporal semantics for a language with channel-based process communication.

A compositional temporal semantics, based on the style of Barringer *et al.* (1985), has been attempted for an Ada tasking subset in the project report of Burton (1986).

4   SOME TROUBLE WITH ABSTRACTION

4.1   The Problem

Previous applications of temporal logic as a language for the semantic description of (concurrent) program languages, both in a *global* sense, see Manna and Pnueli (1984), and in the more recent compositional or denotational style as exemplified in the above sections, have been based upon a *linear* and *discrete* time model. In essence, a temporal formula (schema) is given for each type of language construct such that the set of models of that formula represent, in some obvious way, the possible behaviours of the particular construct. Considering just nonconcurrent programming languages, a linear-time model is used because linear execution traces (or histories) are a most natural (and quite operational) model of program execution behaviour. Furthermore, the time model is chosen to be discrete because, again, a discrete state-change semantics is a most natural (operational) view of low-level machine execution. Now as concurrent program behaviour can be easily modelled by all possible interleavings of the discrete, linear, execution sequences arising from the separate 'sequential' processes of the concurrent program (interleaving semantics), linear discrete temporal logic is clearly an appropriate tool.

There are, however, several criticisms that can be (and, indeed, are) made about the use of such a logic in semantic descriptions. The major point is that this form of temporal logic forces too much irrelevant detail to be present in the semantic descriptions; it appears that the lowest level of atomicity is forced to be visible. This is, of course, quite unnecessary for descriptions of *interference-free* transformational components where a relational semantics (input-output semantics) is quite sufficient. On the other hand, for concurrent program behaviour with interaction through shared variables, it is necessary that certain intermediate states (i.e. states where shared variables are altered) are visible. However, the above-mentioned uses of temporal logic ensure that not only are the desired global state changes visible, but also the ticks of other local, supposedly invisible, state changes. Similar criticism applies also in the use of linear discrete temporal logic in semantic descriptions of parallel programming languages where communication occurs through message-passing. Many critics lay the blame on the use of the next-time operator; we certainly agree this does appear to cause some trouble, but its removal is not sufficient, in our view, to obtain *good* temporal semantics.

The upshot of the above remarks is that linear discrete temporal logic will not usually be able to provide *fully abstract* semantics with respect to some desired level of observation.

Below, the above discussion is amplified by a concrete example. Some formal notions of what is desired then follow and finally an indication of possible approaches to obtaining temporal languages capable of providing abstract semantics is given.

4.2 Example

Consider again the distributed-variables programming language and its open temporal semantic description. In particular, consider the temporal description of assignment. Assuming the *owned* variables of the process are $\bar{y}$, the semantics is of the form:

$$\mathcal{F}_\Theta(x := e; S) \Leftrightarrow I(\bar{y})\mathcal{U}(\bigcirc(x = e) \land I(\bar{y} - \{x\}) \land \bigcirc\mathcal{F}_\Theta(S))$$

To simplify matters, consider the semantic description of assignment when the only owned variable is x. The description can be rewritten as below.

$$\mathcal{F}_\Theta(x := e; S) \Leftrightarrow \exists u . x = u \land \bigcirc(x = u\mathcal{U}(x = e \land \mathcal{F}_\Theta(S)))$$

This particular form emphasizes the fact that the given semantics of assignment enforces at least one *tick* corresponding to the *atomic* execution of the assignment. A formula $\phi\mathcal{U}\psi$ is satisfied by any model (σ, α, i) that satisfies the formula ψ, and thus $\theta \land \bigcirc(\phi\mathcal{U}\psi)$ is satisfied by any model σ, α, i that satisfies θ and $\bigcirc\psi$.

Now consider the semantics for the program fragment $x := 1; \ x := 2; \ S$ running again within a process which owns just the variable x. For convenience, we let $\phi\mathcal{U}^\oplus\psi$ denote $\phi \land \bigcirc(\phi\mathcal{U}\psi)$, and let the resulting language be known as $\mathcal{L}^\oplus$. We then obtain a formula equivalent to the following.

$$\exists u . x = u\mathcal{U}^\oplus(x = 1\mathcal{U}^\oplus(x = 2 \land \phi))$$

where ϕ denotes the temporal formula for the 'semantics' of S. Note that this semantics forces at least *two* ticks corresponding to the atomic execution of the statements $x := 1$ and $x := 2$. This can cause a problem. Consider now the semantics obtained for $x := 1; \ x := x; \ x := 2; \ S$ in the same environment.

$$\exists u . x = u\mathcal{U}^\oplus(x = 1\mathcal{U}^\oplus(x = 1\mathcal{U}^\oplus(x = 2 \land \phi)))$$

Clearly, this semantics enforces at least three ticks (seen by the three occurrences of $\mathcal{U}^\oplus$). But, from an observational point of view where all we can see are state changes, the null effect of $x := x$ should not be visible. In fact we would like to obtain the same, or at least an equivalent, semantic temporal formula as in the first case. Such troubles have, of course, been indicated before, and it has been suggested that the next-time operator should not be used, as in Lamport (1983a, b). Thus, defining $\phi\mathcal{U}^+\psi$ as the formula

$\phi \wedge (\phi \mathcal{U} \psi)$, and referring to the logic as $\mathcal{L}^+$, what is then suggested as the semantics for the above cases is as below.

$$\exists u . x = u \mathcal{U}^+ (x = 1 \mathcal{U}^+ (x = 2 \wedge \phi))$$

$$\exists u . x = u \mathcal{U}^+ (x = 1 \mathcal{U}^+ (x = 1 \mathcal{U}^+ (x = 2 \wedge \phi)))$$

Now as $x = 1 \mathcal{U}^+ (x = 1 \mathcal{U}^+ \psi)$ is equivalent to $x = 1 \mathcal{U}^+ \psi$, of course, it is trivial to see that these are equivalent formulae as desired.

The moral of this first example is that the semantic description should be *insensitive to finite stuttering*.

As a second simple example consider the semantic descriptions for a call of the procedure P which is defined as $P ::= x := x; P$. In the semantics without next-time, we look for the (maximal) solution to

$$\phi \Leftrightarrow \exists u . x = u \mathcal{U}^+ \phi$$

which is clearly the temporal formula *true*. Unfortunately, *true* represents chaotic behaviour and is not what we expect. The behaviour of P should clearly be that x remains constant forever. We say that this semantics is *insensitive* to *infinite stuttering*.

The semantic description obtained using the temporal logic with the next-time operator gives the correct behaviour and is *sensitive* to *infinite stuttering*. We look for the solution to

$$\phi \Leftrightarrow \exists u . x = u \mathcal{U}^\oplus \phi$$

which is $\exists u . \Box x = u$, because a 'next' is forced on every iteration, and thus $x = u$ is extended for all time.

Thus, there is a dilemma. The language without next-time is fine for just finite programs, but gives wrong (intuitively) semantics for the silent divergence, whereas the language with next-time is fine for silent divergence but wrong for the finite case.

A logic that will enable semantic descriptions which are

INSENSITIVE to FINITE STUTTERING

and

SENSITIVE to INFINITE STUTTERING.

is thus desired.

4.3 Some Formal Notions

Here the notions of abstractness and expressiveness that are used are set out somewhat more formally.

In order to claim that some 'semantic' description of a language is, indeed, the semantics of that language, that purported semantics must be judged against an accepted 'natural' semantics. Usually this natural semantics is *operational*, often close to some implementation of the language. Such a natural operational semantics will be denoted by $\mathscr{A}$. The operational aspects can (depending on the complexity of the language) cause the description to be cluttered with unnecessary detail and hence make the description unusable as a 'proper' semantics. Typically, one may find that programs which are accepted as being 'identical' in a certain sense have different semantic descriptions (cf. the example of the previous section).

When faced with an object that possesses too much detail, it is usual to define some abstraction mechanism which hides away irrelevances. In this treatment of semantics, such abstraction occurs through an observation function $\mathcal{O}$ which filters out aspects that should not be seen. It might thus appear that the composition $\mathcal{O}.\mathscr{A}$ provides the proper abstract semantics. Unfortunately, life is never that straightforward! The above composition yields, essentially, a *trace* semantics. It is, of course, well known that trace semantics are not usually strong enough. Consider a simple language with synchronous channel communication. Taking observations to be just the communication histories, a program which deadlocks after a communication trace t will not necessarily be distinguished from one which terminates after trace t.

However, now armed with the fixed entities $\mathscr{A}$ and $\mathcal{O}$, it is possible to set up *the* semantics of the language. It is required that *equivalent* programs are substitutive, i.e. given two equivalent programs (or parts thereof) they ought to be fully interchangeable. Put another way, no program *context* should be able to distinguish two equivalent programs; the equivalence relation should thus be *a congruence*.

A semantics given by $\mathscr{M}$ is defined to be *expressive* with respect to the semantics, induced under contexts, C, by $\mathcal{O}.\mathscr{A}$, if:

$$\forall P_1, P_2 . \mathscr{M}(P_1) = \mathscr{M}(P_2) \Rightarrow \forall C . \mathcal{O}.\mathscr{A}(C[P_1]) = \mathcal{O}.\mathscr{A}(C[P_2])$$

Furthermore, a semantics given by $\mathscr{M}$ is *abstract* with respect to the semantics induced under contexts by $\mathcal{O}.\mathscr{A}$ if:

$$\forall P_1, P_2 . (\forall C . \mathcal{O}.\mathscr{A}(C[P_1]) = \mathcal{O}.\mathscr{A}(C[P_2])) \Rightarrow \mathscr{M}(P_1) = \mathscr{M}(P_2)$$

Finally, a semantics given by $\mathscr{M}$ is *fully abstract* with respect to the semantics induced under contexts by $\mathcal{O}.\mathscr{A}$ if it is both expressive and abstract.

In providing a temporal semantics for a programming language, the function $\mathscr{F}$ translates a language construct to an appropriate temporal logical formula. The semantics of some language construct S is then the set of formulae equivalent to the formula given by $\mathscr{F}$, i.e.

$$\{\phi \mid \vDash \phi \Leftrightarrow \mathscr{F}_\Theta(S)\}$$

Note that the set of models satisfying $\mathscr{F}_\Theta(S)$ would be just as good.

Thus, in order to achieve fully abstract temporal semantics, an isomorphism between the equivalence classes induced by $\mathscr{F}$ and the equivalence classes induced by $\mathcal{O}\mathscr{A}$ under contexts must be obtained. These relationships are depicted in Fig. 4.

4.4 Possible Approaches to Abstractness

Consider Fig. 4 in the case of the logic $\mathscr{L}^+$ and the distributed-variables language. Full abstractness of the temporal semantics failed because too many programs were identified to be equivalent, for example silent divergence was identified with the chaotic process. How can things be altered to obtain the 'right' number of equivalences? Given that $\mathcal{O}$, the observation function (on the operational semantics), and $\mathscr{A}$, the operational semantic function, are fixed, one is at liberty to change $\mathscr{F}$, $\mathscr{T}\mathscr{L}$ language or $\vDash$. Assuming that change to $\mathscr{F}$ will not achieve the desired results, as we believe to be the case, change for the good should come from modifying either the

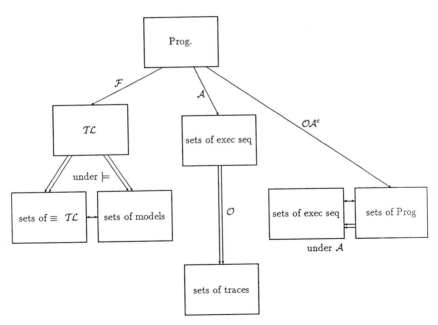

Fig. 4. Semantic relationships.

logical language, say by introducing new operators, or by changing the models and interpretation of the logical language. (Note, of course, that a change in logical language does imply some change to the 'semantic' function $\mathcal{F}$). In the subsections below, three approaches are briefly outlined. The first changes the language, the second and third change the language and its models.

4.4.1 Collapse of the expanding state

With careful consideration of the failure of $\mathcal{L}^+$ it becomes clear that the difficulty came about through a nasty case of unguarded recursion, which then gives rise to chaotic behaviour in the infinite case. Clearly, in the model, the semantics of recursion can be forced to be always guarded by prefixing the models of all approximations of the recursive behaviour by a copy of their first state. Then, after the solution of the guarded recursive (in)equation has been formed, the 'extra' states can be eliminated as necessary. From the model view, the extra states are not eliminated if the resulting solution is an infinite sequence of identical states.

Unfortunately, linear temporal logic does not possess such sequence-oriented operations. The answer is clearly to create a logic that does! Two new operators, **exp** and **coll**, are added to those of $\mathcal{T}\mathcal{L}$ to mimic the above suggestion. Currently their semantics is given as below.

$$\sigma, \alpha, i \vDash \mathbf{exp}(\phi, \bar{y}, k) \quad \textbf{iff} \quad \begin{cases} \sigma, \alpha, i + 2 \vDash \phi & \text{and} \\ \sigma, \alpha, i & \vDash clock(\bar{y}, k) \end{cases}$$

where

$$clock(\bar{y}, k) \stackrel{\text{def}}{=} (k \wedge \bigcirc \neg k) \wedge \exists \bar{u} . \bar{y} = \bar{u} \wedge \bigcirc \bar{y} = \bar{u} \wedge \bigcirc\bigcirc \bar{y} = \bar{u}$$

So the models satisfying the formula $\mathbf{exp}(\phi, \bar{y}, k)$ are the models of ϕ prefixed by two states preserving the values of the variables $\bar{y}$ but clocking the proposition k.

$$coll((\sigma, \alpha, i), \bar{y}, k) \vDash \mathbf{coll}(\phi, \bar{y}, k) \quad \textbf{iff} \quad \sigma, \alpha, i \vDash \phi$$

where

$$coll((\sigma, \alpha, i), \bar{y}, k) \stackrel{\text{def}}{=}$$

$$\begin{cases} s_i \ldots s_n . coll((\sigma, \alpha, n + 3), \bar{y}, k)_\sigma, \\ a_i \ldots a_n . coll((\sigma, \alpha, n + 3), y, k)_\alpha, & \text{if} \\ 0 \\ \\ \sigma, \alpha, i \end{cases} \quad \begin{cases} \sigma, \alpha, i \vDash \neg \exists \bar{u} . \square \bar{y} = \bar{u} \\ \text{and for some } n \geqslant i \\ \sigma, \alpha, n + 1 \vDash clock(\bar{y}, k) \\ \text{and for all } j, i \leqslant j \leqslant n, \\ \sigma, \alpha, j \nvDash clock(\bar{y}, k) \\ \text{otherwise} \end{cases}$$

And so the models which satisfy $\mathbf{coll}(\phi, \bar{y}, k)$ are collapsed models of ϕ. The reduction process deletes the *unnecessary* guarding or clocking states, presumably introduced by application of the **exp** operator, from the models. In cases where guarding was necessary, e.g. silent divergence, the model is not collapsed.

As yet axiomatization of the new logic has not been considered.

The following illustrates the use of **exp** and **coll** in the semantic description of the distributed variables language. Suitable semantic descriptions can be obtained by first modifying the descriptions of Section 3.1 to use the logic without next-time, as indicated in Section 4.2. The semantics thus obtained will be insensitive to both finite and infinite stuttering. The procedure definition semantics should then be modified as below.

$$\mathscr{F}'\left(p \overset{\text{def}}{=} S_p; S\right) \overset{\text{def}}{=} \mathscr{F}_\Theta(S)$$

where

$$\Theta \overset{\text{def}}{=} [p \to \exists k.\mathbf{coll}(\nu\xi.\mathbf{exp}(\mathscr{F}_{[p \to \xi]}(S_p), \bar{y}, k), \bar{y}, k)]$$

and where $\bar{y}$ are the owned variables of the process.

It is fairly easy to convince oneself that the examples in Section 4.2 now have the right abstract semantics. It is only necessary to consider the semantics of a silently divergent process, e.g. a call of p where $p \overset{\text{def}}{=} p$. The semantics is obtained in two stages. First an artificially expanded behaviour is obtained for p, e.g. $\nu\xi.\mathbf{exp}(\mathscr{F}_{[p \to \xi]}(S_p), \bar{y}, k)$. This behaviour contains a clock tick, preserving the owned variables of p, for every recursive call of p, and thus implies that $\exists \bar{u}.\Box(\bar{y} = \bar{u})$, where $\bar{y}$ are the owned variables of p. The second stage of construction of the semantics removes the clock ticks but only in the case when there is no change to the owned variables. Thus the final semantic formula for p still implies $\exists \bar{u}.\Box(\bar{y} = \bar{u})$. Although no proof has been produced it is reasonably clear that a fully abstract semantics results. What is not so clear is whether a reasonable and practicable axiomatization is feasible.

4.4.2 Next-change logic—EL

In the work of Fisher (1986), a logic (EL) is proposed in which the linear discrete models are restricted by:

$$\forall i \in (|\mathscr{M}| - 1).S_i = S_{i+1} \Rightarrow (\forall j \geqslant i.S_j = S_i)$$

where $\mathscr{M}$ is an EL model, and the S_i are the 'states' in the model. Here, S_{i+1} is used to represent the $i + 1^{st}$ state in the EL model sequence, and two states are equal if exactly the same propositions are true in each state.

This restriction means all 'adjacent' states must be different, except in the case of an infinite sequence of equal states. An infinite sequence is forced if we have two equal adjacent states, otherwise the model may be finite or infinite.

So, essentially, EL is a linear temporal logic with the restriction that we ignore 'repeated' states and model only the states where 'something' (a proposition or variable, depending on what equivalence we are taking) has changed since the last state. The only exception to this compression rule is the case where nothing (no proposition) changes for an infinitely long time, in which case compression to a finite sequence is not required, so that infinite behaviour is still exhibited.

EL is similar in some respects to the temporal logic used by Lamport though the adopted model is changed more radically than in his case and an attempt to abstract from linear temporal logic is made. Motivation has come from the same dissatisfaction with 'standard' temporal logics that Lamport cites. Also the 'next-time' operator is seen as being too concrete, and consequently forcing linear temporal logic to model too much information.

So, although a 'next' operator in EL does exist, it is more like a 'next-change' operator, as it takes the next state in the sequence at which something will have changed (unless an infinite sequence of identical states happens). The basic operators in EL are the ('normal') propositional calculus operators, the 'maximal fixed point' operator, and this 'next-change' operator, **A**.

4.4.3 Dense time

The third approach that has been considered, and the most fully evaluated so far, changes the underlying time structure from being discrete to being dense. The effect of this on a semantic description of the distributed variables language is seen to have two advantages.

1. A fully abstract semantics (with respect to the chosen level of observation, i.e. state changes) is attainable.
   (a) Insensitivity to finite stuttering arises because within some finite interval (of no change) it is possible to 'squash' as many other finite intervals as required. A *finite variability* restriction is applied to the dense time structure; this ensures that a proposition (variable) does not change infinitely within a finite interval and, therefore, allows ω-based induction and other such nice techniques to be valid.
   (b) Sensitivity to infinite stuttering does not come immediately through use of a dense time structure. It is realised by forcing

some unknown global proposition to change value in the semantics of a procedure call. Thus, because of finite variability, if a proposition changes infinitely, then it must be changing over a truly infinite period. (There is some similarity between this and the clocking introduced in the expand and collapse approach.)

2. The dense time model allows the interleaving model of parallelism to be thrown away in favour of real parallelism. After all, for example, five intervals (representing five sequential actions of one process) may occur in parallel (i.e. overlap) with two intervals from some other process.

Experience so far has indicated that working with the temporal language based on dense time is really not harder than with the corresponding discrete-time logic.

Readers interested in a fuller account of the dense time model and its temporal logic should consult Barringer *et al.* (1986).

Acknowledgements

The work presented in this article has grown from the collaborative research with Ruurd Kuiper and Amir Pnueli reported in the BKP papers. The author thanks both Ruurd and Amir for many hours of stimulating discussion enjoyed during several periods of intense collaboration.

The author also wishes to express his thanks to all those who have commented on early drafts of this article, in particular to Antony Galton for his careful and illuminating criticism.

The work reported here has been supported under SERC grant GR/C/05760 and Alvey PRJ/SE/054 (SERC grant GR/D/57942)

References

Banieqbal, B. and Barringer, H. (1986), *A Study of an Extended Temporal Language and a Temporal Fixed Point Calculus*, Technical Report UMCS-86-11-1, University of Manchester, Department of Computer Science.

Barringer, H. and Fisher, M. D. (1987), *Temporal Semantics for Channel-Based Communication*, Technical Report in preparation, University of Manchester, Department of Computer Science.

Barringer, H., Kuiper, R. and Pnueli, A. (1984), 'Now you may compose Temporal Logic specifications', in *Proceedings of the Sixteenth ACM Symposium on the Theory of Computing*, Washington, D.C., 51–63.

Barringer, H., Kuiper, R. and Pnueli, A. (1985), 'A compositional temporal approach to a csp-like language', in E. J. Neuhold and G. Chroust (eds) *Formal Models of Programming* North-Holland: Elsevier, 207–227.

Barringer, H., Kuiper, R. and Pnueli, A. (1986), 'A really abstract concurrent model and its temporal logic', in *Proceedings of the Thirteenth ACM Symposium on the Principles of Programming Languages* Florida: St Petersberg Beach, 173–183.

Burton, M. D. (1986), *A Temporal Semantics for an Ada Tasking Subset*, 3rd Year Project Report, University of Manchester, Department of Computer Science.

Fisher, M. D. (1986), 'Temporal logics for abstract semantics,' PhD thesis, in preparation, University of Manchester, Department of Computer Science.

Hoare, C. A. R. (1969), 'An axiomatic basis for computer programming', *Comm. ACM*, **12**, 576–583.

Jones, C. B. (1980), *Software Development—a Rigorous Approach* New Jersey: Prentice-Hall.

Jones, C. B. (1983), 'Specification and design of (parallel) programs', in R. E. A. Mason (ed.) *Information Processing 83* IFIP, North-Holland: Elsevier Science Publishers BV. 321–332.

Lamport, L. (1983a), 'Specifying concurrent program modules', *ACM Transactions on Programming Languages and Systems*, **5**(2), 190–222.

Lamport, L. (1983b), 'What good is temporal logic', in R. E. A. Mason (ed.) *Information Processing 83* North-Holland: Elsevier, 657–668.

Manna, Z. and Pnueli, A. (1981), 'Verification of concurrent programs: the temporal framework', in Robert S. Boyer and J. Strother Moore (eds) *The Correctness Problem in Computer Science* London: Academic Press.

Manna, Z. and Pnueli, A. (1984), 'Adequate proof principles for invariance and liveness properties of concurrent programs', *Science of Computer Programming*, **4**, 257–289.

Milner, R. (1980), *A Calculus of Communicating Systems*, Volume 92 of *Lecture Notes in Computer Science* Berlin: Springer-Verlag.

Owicki, S. S. and Gries, D. (1976a), 'An axiomatic proof technique for parallel processes', *Acta Informatica*, **6**, 319–340.

Owicki, S. S. and Gries, D. (1976b), 'Verifying properties of parallel programs: an axiomatic approach'. *Comm. ACM*, **19**(5), 279–285.

Owicki, S. and Lamport, L. (1982), 'Proving liveness properties of concurrent programs', *ACM Transactions on Programming Languages and Systems*, **4**(3), 455–495.

de Roever, W. P., Jr (1985), 'The quest for compositionality—a survey of assertion-based proof systems for concurrent programs. Part 1: concurrency based on shared variables', in E. J. Neuhold and G. Chroust (eds) *Formal Models of Programming* North-Holland: Elsevier, 181–205.

Wolper, P. (1983), 'Temporal logic can be more expressive', *Information and Control*, **56**.

3 TEMPORAL LOGIC PROGRAMMING

Roger Hale

Computer Laboratory, University of Cambridge

1 INTRODUCTION

A serious drawback of conventional logic programming languages is their inability to describe changing systems. They deal only with static descriptions. In fact, Kowalski (1979) argues that the part of an algorithm which is concerned with change is somehow outside the scope of logic. In this chapter we take a rather different view. Our principal concern is with mechanism, and for this purpose, we argue, Temporal Logic is a more suitable starting point for a programming language than either Horn Clause Logic or the more conventional basis of *Pascal* and its derivatives. A brief example may illustrate the point.

Suppose one wishes to instruct a computer in the solution to the *Towers of Hanoi* problem. It is simple enough to specify the goal in a language such as *Prolog*: All the rings from one peg must be moved to another peg. However, it is the mechanism which is all-important. Certain rules must be followed in moving the rings. This is harder to describe in Horn Clause Logic, and although the problem can easily be solved in an algorithmic language such as *Pascal*, to reason about that solution soon becomes cumbersome. We shall return to this example in a later section.

In this paper we do not use classical linear-time temporal logic, but rather Moszkowski's *Interval Temporal Logic* (ITL), whose unit of consideration is the interval as opposed to the individual state (Moszkowski, 1983). ITL is described in the next section. There then follows a brief account of how ITL may be given an operational interpretation. The main body of this chapter comprises some examples to illustrate how this framework may be applied. The examples are all concerned with the representation of real physical systems, ranging from animation to data transmission. They show how different programming paradigms fit into the Temporal Logic framework. These include conventional *Pascal*-like algorithms as well as concurrency, timing, and hierarchical specification.

Temporal Logics and their Applications

ISBN: 0-12-274060-2

2   INTERVAL TEMPORAL LOGIC

Interval Temporal Logic is a modal logic, and therefore extends classical logic with operators to express possibility and necessity. These new operators relate statements about intervals to statements about their subintervals. Thus the necessity operator ($\Box$) is to be understood to mean: 'on all subintervals', and the possibility operator ($\Diamond$) to mean: 'on at least one subinterval'. The logic is further extended with the powerful *semicolon* operator (;) which denotes sequential composition, and the *next* operator ($\bigcirc$) which refers to the next subinterval (the tail of the current interval).

In the remainder of this section we shall give a review of ITL, including formal definitions of the operators mentioned above. Familiarity with classical logic is assumed, but the following account should be accessible to those with no prior knowledge of either modal or temporal logic. In case of difficulty, refer to Chellas (1980) for a good introduction to modal logic, or to Rescher and Urquhart (1971) for the definitive text on temporal logic. The reader who is deterred by mathematical definitions without much intuitive motivation should perhaps defer this section to a second reading.

2.1   Semantics

In order to describe the logic formally we need to know how to construct formulae and how to interpret them. Syntactically, any formula of classical predicate logic is also a formula of ITL. Furthermore, any ITL formula w may be prefixed by a modal operator to produce a new formula $\Box w$, $\Diamond w$, or $\bigcirc w$; and any two formulae, w_1 and w_2, may be combined sequentially in the formula $w_1; w_2$. Semantically, we model ITL in terms of:

A universe of intervals, Σ,
A subinterval relationship, R, and
A meaning function, M

Each interval $\sigma \in \Sigma$ is a non-empty sequence of states: $\sigma = \langle \sigma_0 \ldots \sigma_{|\sigma|} \rangle$, where the length $|\sigma|$ of the interval is the number of *transitions* between states. We shall assume in the following that intervals are temporal, which means that each state corresponds to a moment in time. It is, in fact, a function which assigns a value to each variable of the system.

In contrast, the value of a variable on an *interval* is determined by the meaning function. We distinguish two kinds of variables: *state* variables and *static* variables. The value of a state variable on an interval is just its value on the first state of that interval; and the value of a static variable is constant on a given interval, but may change from interval to interval. State variables are conventionally denoted by names starting with a capital letter, static variables by lower case names. For the state variable A:

$$M_\sigma[A] = \sigma_0(A).$$

An equivalent formula holds for a static variable a, but in addition we have

$$\sigma_i(a) = \sigma_j(a) \quad \text{for all} \quad 0 \leqslant i, j \leqslant |\sigma|.$$

The meaning function also assigns a truth value to each formula on each interval, and so furnishes an interpretation of formulae. An interval σ *satisfies* a formula w in a given interpretation if w is true on σ in that interpretation. We write:

$$\sigma \vDash w \quad \text{iff} \quad M_\sigma[w] = true.$$

For example, the formula $A = 0$ is true on an interval σ if the value of A is 0 on the first state.

$$\sigma \vDash A = 0 \quad \text{iff} \quad \sigma_0(A) = 0.$$

A *valid* formula is true on every interval in Σ. In recognition of this fact we omit the satisfying interval.

$$\vDash (A = 0) \vee (A \neq 0).$$

2.2 The Operators $\square$ and $\diamondsuit$

There are three subinterval relationships, R_a, R_i and R_t, to express the notions 'any-', 'initial-' and 'terminal subinterval', respectively.

$$\sigma' R_a \sigma \quad \text{iff} \quad \sigma' \text{ is a subsequence of } \sigma.$$

$$\sigma' R_i \sigma \quad \text{iff} \quad \sigma' \text{ is an initial subsequence of } \sigma.$$

$$\sigma' R_t \sigma \quad \text{iff} \quad \sigma' \text{ is a terminal subsequence of } \sigma.$$

Consequently, three different pairs of operators may be defined corresponding to the different varieties of subinterval. For a formula w:

$$\sigma \vDash \diamondsuit w \quad \text{iff} \quad \text{there is an interval } \sigma' \text{ such that } \sigma' R_x \sigma \text{ and } \sigma' \vDash w,$$

$$\sigma \vDash \boxed{\times} w \quad \text{iff} \quad \text{for every interval } \sigma' \text{ such that } \sigma' R_x \sigma, \sigma' \vDash w,$$

where x may be 'a', 'i' or 't'. Thus, the following statements are true for the three-state interval $\langle stu \rangle$:

$$\langle stu \rangle \vDash \diamondsuit w \quad \text{if} \quad \langle stu \rangle \vDash w,$$

$$\langle stu \rangle \vDash \boxed{t} w \quad \text{iff} \quad \langle stu \rangle \vDash w \text{ and } \langle tu \rangle \vDash w \text{ and } \langle u \rangle \vDash w.$$

We may also infer from our definitions that:

$$\boxed{\times}w \equiv \neg \diamondsuit \neg w,$$

$$\boxed{\times}w \rightarrow w,$$

$$\boxed{\times}w \rightarrow \boxed{\times}\,\boxed{\times}w.$$

The first sentence expresses the so-called *duality* of $\square$ and $\diamondsuit$, and holds in all classical systems of modal logic. The other two characterize a modal logic of type S4. They are valid because each of the relations, R_a, R_i and R_t, is reflexive and transitive.

It will become clear as we proceed that the terminal forms of the operators $\square$ and $\diamondsuit$ assume a dominant role in the application of ITL, the other forms being somewhat less useful. For this reason we frequently omit the annotations from the terminal operators.

2.3 Sequential Composition

Sequential composition is expressed by the *semicolon* or *chop* operator. The composition of two formulae $w_1; w_2$ is defined to be true on an interval σ if it is possible to divide σ into two subintervals in a way that makes w_1 true on the first and w_2 true on the second. The definition employs an auxiliary operator $\odot$ which composes two *intervals* sequentially.

$$\sigma \vDash w_1; w_2 \quad \text{iff} \quad \text{there are intervals } \sigma' \text{ and } \sigma''$$

$$\text{such that } \sigma = \sigma' \odot \sigma''$$

$$\text{and } \sigma' \vDash w_1$$

$$\text{and } \sigma'' \vDash w_2,$$

where

$$\sigma = \sigma' \odot \sigma'' \quad \text{iff} \quad \sigma' R_i \sigma \text{ and } \sigma'' R_t \sigma \text{ and } |\sigma'| + |\sigma''| = |\sigma|.$$

Note that the two intervals share a common state. For example:

$$\langle \text{stuv} \rangle \vDash w_1; w_2 \quad \text{if} \quad \langle \text{stu} \rangle \vDash w_1 \quad \text{and} \quad \langle \text{uv} \rangle \vDash w_2.$$

Sequential composition is a powerful concept. In particular, it can be used to express the operators $\square$ and $\diamondsuit$.

$$\boxed{\diamondsuit}w \equiv \textit{true}; w; \textit{true},$$

$$\boxed{\diamondsuit}w \equiv w; \textit{true},$$

$$\boxed{\diamondsuit}w \equiv \textit{true}; w.$$

The corresponding necessity operators can be defined by duality. Therefore, these operators may be viewed as derived, and not essential to the logic.

2.4 Next and Other Operators

As the intervals in Σ are discrete, the concept of a *next* subinterval is well-defined. This is expressed by the operator $\bigcirc$. The formula $\bigcirc w$ holds on an interval of length one or more if w holds on the next terminal subinterval.

$$\sigma \vDash \bigcirc w \quad \text{iff} \quad \text{there is an interval } \sigma'$$

$$\text{such that } \sigma' R_t \sigma$$

$$\text{and } \sigma' \vDash w$$

$$\text{and } |\sigma'| = |\sigma| - 1.$$

This operator can also be applied to an expression, yielding its value on the next terminal subinterval.

Finally, because it extends classical predicate logic, ITL contains all the familiar operators, including universal and existential quantifiers. Be aware, though, that in the temporal world their intuitive meaning may be different. Conjunction and disjunction ($\wedge$ and $\vee$), for example, assume the semantics of parallel composition and non-determinism, respectively.

$$\sigma \vDash w_1 \wedge w_2 \quad \text{iff} \quad \sigma \vDash w_1 \text{ and } \sigma \vDash w_2 \text{ concurrently.}$$

3 TEMPORAL LOGIC PROGRAMMING

Just as a formula of Horn Clause Logic can be given an operational interpretation in the programming language *Prolog* (Clocksin and Mellish, 1981), so certain formulae of ITL may be regarded as programs in a language called *Tempura*. The methods of execution in the two cases are quite different, however. Where *Prolog* regards programs as problems to be solved by use of the Resolution Principle, *Tempura* programs are executed by transformation.

3.1 Program Execution

In the process of execution, a *Tempura* program w is reduced to a part which specifies the current state, and a part which is true on the next terminal subinterval. As the current state advances, so this reduction is repeated until finally the current subinterval is just a single state and there is no next subinterval. The reduction is then complete, and w is reduced to a conjunction of subformulae:

$$w \equiv \bigwedge_{i=0}^{n} \bigcirc^i w_i,$$

where the notation $\bigcirc^i w$ indicates i applications of the operator $\bigcirc$ to the formula w (or zero, if $i < 0$). Each subformula w_i specifies a state σ_i which is the first state of the corresponding subinterval, and together these states make up an interval $\langle \sigma_0 \dots \sigma_n \rangle$ which satisfies the original program.

As a simple illustration, consider the term $\square (A = 1)$ on an interval of four states. The reduction process generates the following conjunction:

$$(A = 1) \wedge \bigcirc(A = 1) \wedge \bigcirc\bigcirc(A = 1) \wedge \bigcirc\bigcirc\bigcirc(A = 1).$$

This is not a complete *Tempura* program because the original and transformed formulae are not equivalent. The original does not specify an interval length, whereas the transformation may only hold on an interval of four or more states.

Our reduction technique does not work for arbitrary formulae because not every formula can be reduced to a conjunction of the above form. To admit disjunction into the reduced form would increase considerably the complexity of the execution algorithm, since in general one must use back-tracking to generate all intervals satisfying the alternate formula $w_1 \vee w_2$. To handle the negated formula, $\sim w$, one must in addition assume a 'closed world', as *Prolog* does. Furthermore, an exhaustive search of the interval space is much more costly than the *Prolog* equivalent because of the additional temporal dimension. For this reason the set of executable formulae is restricted to those adjudged to be 'efficiently manageable', and neither alternate, nor negated formulae are included. This judgement is somewhat arbitrary. Indeed, the version of *Tempura* we use is more general than the one described by Moszkowski (1986), but in any event the set of *Tempura* programs must be a subset of ITL.

3.2 New Operators

There is a natural correspondence between certain logical operators and certain aspects of programming. For example, the conjunction and semicolon operators model parallel and sequential composition, and existential quantification corresponds to the introduction of local variables. In a formula such as the following, for example, there are two instances of a variable I with different scopes:

$$\exists I : \{I = 0 \wedge \dots \wedge \exists I : \{I = 1 \wedge \dots\} \wedge \dots\}$$

Likewise, bounded universal quantification instantiates a number of concurrent copies of a formula, as in the following example:

$$\forall i < n : p(i) \equiv p(0) \wedge \ldots \wedge p(n - 1)$$

Other concepts, such as unit delay and assignment, have no direct analogue, but may easily be defined in terms of the basic operators. Consequently, a body of new operators has arisen to simplify the expression of these ideas. Some of the most useful derived operators are defined below.

In the following definitions w denotes an arbitrary formula, n stands for an integer expression, x is a static (unchanging) value, and A, B and C are state (changing) variables. More generally, A, B and C might be replaced by expressions.

empty $\equiv_{\text{def}} \neg \bigcirc true$ (the zero-length interval)

skip $\equiv_{\text{def}} \bigcirc$ *empty*   (the unit-length interval)

len n $\equiv_{\text{def}} \bigcirc^{n}$ *empty*   (interval length)

keep w $\equiv_{\text{def}} \Box(\neg empty \rightarrow w)$ (on all but the last state)

A gets B $\equiv_{\text{def}}$ *keep* $((\bigcirc A) = B)$ (unit delay)

stable A $\equiv_{\text{def}}$ *A gets A*   (constant value)

halt C $\equiv_{\text{def}} \Box(C \equiv empty)$ (interval termination)

fin w $\equiv_{\text{def}} \Box(empty \rightarrow w)$ (on the last state)

A ← B $\equiv_{\text{def}} \exists x : (x = B \wedge fin\ A = x)$ (assignment)

if C then w_1 else w_2 $\equiv_{\text{def}} (C \rightarrow w_1) \wedge (\neg C \rightarrow w_2)$ (conditional)

while C do w $\equiv_{\text{def}}$ *if* $\neg C$ *then empty*
            else (w; while C do w)   (repetition)

Different kinds of repetitive constructs, such as *for* loops, can be defined in a way analogous to the *while* loop.

Let us give an example to make these definitions concrete. Shown below is the behaviour a system of variables A, B and C on an interval $\langle ststsu \rangle$. The system alternates between states s and t before entering its final state u.

State	s	t	s	t	s	u
A	0	0	0	0	0	1
B	1	0	1	0	1	0
C	1	1	1	1	1	1

The following statements are true of this system:

$$\langle u \rangle \vDash \textit{empty},$$

$$\langle st \rangle \vDash \textit{skip},$$

$$\langle ststsu \rangle \vDash \textit{len } 5,$$

$$\langle ststsu \rangle \vDash \textit{keep } (A = 0),$$

$$\langle ststsu \rangle \vDash B \textit{ gets } \neg B,$$

$$\langle ststsu \rangle \vDash \textit{stable } C,$$

$$\langle ststsu \rangle \vDash \textit{halt } (A = 1),$$

$$\langle ststsu \rangle \vDash \textit{fin } (B = 0),$$

$$\langle ststsu \rangle \vDash B \leftarrow A,$$

$$\langle ststsu \rangle \vDash \textit{while } (A \neq 1) \textit{ do } (\textit{skip} \wedge (B \leftarrow \neg B)).$$

It is apparent from these examples that *Tempura* embodies ideas from conventional imperative programming as well as some which are more akin to dataflow. The term $A \leftarrow B$ represents a single destructive assignment, whereas a term like A *gets* B defines a whole sequence of values for A.

4   APPLICATIONS

In this section we give weight to our advocation of Temporal Logic Programming by demonstrating its application to four problems. We begin with a problem of classical simplicity.

4.1   Towers of Hanoi

The *Towers of Hanoi* is a game played with a number of rings, no two of which are the same size. The rings fit onto pegs to form towers, and these towers must be moved between three pegs (*left, centre* and *right*) according to two rules. The first rule is that only one ring may be moved at a time, and the second that no ring may ever rest on one smaller than itself. Initially all the rings are on the *left* peg. The question is: How long does it take to move a tower of *n* rings from the *left* peg to the *centre* peg?

There is an elegant and well-known recursive solution which is gained by observing that a tower of *n* rings can be moved from the *left* peg to the *centre* peg by first moving the top *n-1* rings from the *left* to the *right* peg, then moving the single remaining ring onto the *centre* peg, and finally moving the tower of *n-1* rings from the *right* to the *centre* peg. This process is illustrated for a four-ring tower in Fig. 1.

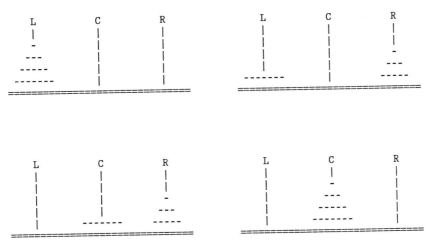

Fig. 1. Moving a tower of four rings.

It is easy enough to formulate this solution in *Tempura*, and a program is shown in Fig. 2. Each peg is represented as a list of integers whose values correspond to the sizes of the rings on that peg. The recursive procedure $move_r(i, A, B, C)$ is used to transfer a tower of i rings from peg A to peg B; and each move of a single ring from peg A to peg B is performed in unit time by the procedure *move-step*(A, B, C). (Notice that it is necessary to keep the peg C stable during the move.)

The reader might well have observed that there is nothing about this program which could not equally well have been done in any language that supports recursion (except perhaps for the parallel execution of program and output). Syntactically, there is little to choose between the *Tempura* solution and the *Pascal* program sketched in Fig. 3. The difference lies in the semantics.

The *Tempura* program is also a precise specification in ITL of an interval which satisfies the *Towers of Hanoi* problem. One can therefore show directly that this program solves the problem in $2^n - 1$ moves. Formally, the program $p_r(n, L, C, R)$ implies the specification $s(L, C, R)$ and an interval length of $2^n - 1$:

$$\vDash p_r(n, L, C, R) \rightarrow s(L, C, R) \wedge \textbf{\textit{len}}(2^n - 1),$$

where

$$p_r(n, L, C, R) \equiv_{\text{def}} L = \langle i{:}0 \leqslant i < n \rangle \wedge C = \langle \rangle \wedge move_r(n, L, C, R),$$

and

$$s(L, C, R) \equiv_{\text{def}} (C \leftarrow L) \wedge \square ordered(L) \wedge \square ordered(C) \wedge \square ordered(R).$$

```
/* Tempura solution to the "Towers of Hanoi" problem */

define hanoi(n) = exists L,C,R : {
  L=[0..n-1] and              /* list from 0 to n-1 */
  C=[] and
  R=[] and
  move_r(n,L,C,R) and
  always display(L,C,R)
}.

/* Move n rings from peg A to peg B */
define move_r(n,A,B,C) = {
  if n=0 then empty
  else {
    move_r(n-1,A,C,B);
    move_step(A,B,C);
    move_r(n-1,C,B,A)
  }
}.

/* Move the topmost ring from peg A to peg B */
define move_step(A,B,C) = {
  skip and
  A <- tail(A) and
  B <- append([head(A)],B) and
  C <- C
}.
```

Fig. 2. A Tempura *solution to the* Towers of Hanoi *problem.*

The predicate *ordered* is true if its argument is an ordered list. Figure 4 shows a sample output of the *hanoi* program which may be seen to satisfy the above properties, a fact that may be verified by executing the specification in parallel with the program. Note also that program output has a semantic value in temporal logic; it is not a side-effect.

With a *Pascal* program on the other hand, one must look elsewhere for formal tools, since neither its correctness, nor its timing behaviour can even be expressed in *Pascal*. The axiomatic system of Hoare (1969) might permit a correctness proof, and the timing might be discovered by solving a recursion equation (for example), but neither can be deduced from the program alone.

We should also point out that a recursive solution in Horn clause form ('pure' *Prolog*) is rather complex, because it must maintain the state history of the system explicitly. The *Prolog* solution shown in Fig. 5 is quite simple, but relies on the extra-logical features which are present in most *Prolog* systems. In particular, it is necessary that clauses are evaluated from left to right for the recursive calls to proceed in the correct order. It is also clear that the *write* statement and the *cut* (!) are present only for their side-effects, since logically

{Pascal solution to the "Towers of Hanoi" problem}

procedure hanoi(n:integer);

```
{Move the topmost ring from peg p1 to peg p2}
procedure move_step(p1,p2:pegtype);
  begin
  ...
  end {of move_step}

{Move k rings from peg p1 to peg p2}
procedure move(k:integer; p1,p2,p3:pegtype);
  begin
  if k=0 then {do nothing}
  else
     begin
     move(k-1,p1,p3,p2);
     move_step(p1,p2);
     move(k-1,p3,p2,p1);
     end
  end {of move}

begin
...
move(n,left,centre,right);
end {of hanoi};
```

Fig. 3. A Pascal *solution to the* Towers of Hanoi *problem.*

PC-Tempura (Version 2.00)

```
Tempura 1> load "hanoi".
[Reading file hanoi.t]
Tempura 2> run hanoi(4).
State   0: L=[0,1,2,3]  C=[ ]        R=[ ]
State   1: L=[1,2,3]    C=[ ]        R=[0]
State   2: L=[2,3]      C=[1]        R=[0]
State   3: L=[2,3]      C=[0,1]      R=[ ]
State   4: L=[3]        C=[0,1]      R=[2]
State   5: L=[0,3]      C=[1]        R=[2]
State   6: L=[0,3]      C=[ ]        R=[1,2]
State   7: L=[3]        C=[ ]        R=[0,1,2]
State   8: L=[ ]        C=[3]        R=[0,1,2]
State   9: L=[ ]        C=[0,3]      R=[1,2]
State  10: L=[1]        C=[0,3]      R=[2]
State  11: L=[0,1]      C=[3]        R=[2]
State  12: L=[0,1]      C=[2,3]      R=[ ]
State  13: L=[1]        C=[2,3]      R=[0]
State  14: L=[ ]        C=[1,2,3]    R=[0]
State  15: L=[ ]        C=[0,1,2,3]  R=[ ]
```

Done! Computation length: 15. Total Passes: 16.
Tempura 3>

Fig. 4. Output of the Tempura *program for a tower of four rings.*

```
/* Prolog solution to the "Towers of Hanoi" problem */

hanoi(N) :- move(N,left,centre,right).

/* Move N rings from peg A to peg B */
move(0,_,_,_) :- !
move(N,A,B,C) :-
  M is N-1,
  move(M,A,C,B),
  move_step(A,B),
  move(M,C,B,A).

/* Move the topmost ring from peg A to peg B */
move_step(X,Y) :- write([move,a,ring,from,X,to,Y]),nl.
```

Fig. 5. A Prolog *solution to the* Towers of Hanoi *problem.*

they are redundant. Therefore we claim that it is no easier to reason about this program than about the *Pascal* one, but we suspect it is the solution which most *Prolog* programmers would produce.

A trickier problem arises if we wish to compare alternate solutions. Suppose that (for pragmatic reasons) we would prefer a non-recursive program. By studying Fig. 4 we may observe that the smallest ring is moved in the same direction (cyclically) on every alternate time step, and that on each intermediate step only one ring other than the smallest can legally be moved, until finally all the rings are on the same peg. Encouraged by this, we propose the *Tempura* program of Fig. 6.

This program uses the iterative procedure $move_i(n, A, B, C)$ to perform the moves. The smallest ring is moved from A to B to C to A, and so on repeatedly, the direction of movement depending on the parity of n. Happily, this program also meets the specification:

$$\vdash p_i(n, L, C, R) \rightarrow s(L, C, R) \wedge \textbf{len}(2^n - 1),$$

where

$$p_i(n, L, C, R) \equiv_{\text{def}} L = \langle i : 0 \leqslant i < n \rangle \wedge C = \langle \rangle \wedge R = \langle \rangle$$

$$\wedge \textbf{\textit{if}} \, odd(n) \, \textbf{\textit{then}} \, move_i(n, L, C, R) \, \textbf{\textit{else}} \, move_i(n, L, R, C).$$

Indeed, the programs p_i and p_r are equivalent:

$$\vdash p_i(n, L, C, R) \equiv p_r(n, L, C, R)$$

How to define this equivalence for the corresponding *Pascal* programs is not so clear.

```
/* Iterative solution to the "Towers of Hanoi" problem */

define hanoi(n) = exists L,C,R : {
   len (2**n-1) and
   L=[0..n-1] and          /* list from 0 to n-1 */
   C=[] and
   R=[] and
   if odd(n) then move_i(n,L,C,R) else move_i(n,L,R,C) and
   always display(L,C,R)
}.

/* Move a tower of n rings */
/* The smallest ring moves from A to B to C to A ... */
define move_i(n,A,B,C) = {
   if n=0 then empty
   else {
      move_step(A,B,C);
      while not(|A|=n or |B|=n or |C|=n) do {
         if smallest_on(A) then move_pair(A,B,C)
         else if smallest_on(B) then move_pair(B,C,A)
         else move_pair(C,A,B)
      }
   }
}.

/* Move the largest ring possible, then the smallest */
/* The smallest ring is on A */
define move_pair(A,B,C) = {
   if legal(B,C) then move_step(B,C,A) else move_step(C,B,A);
   move_step(A,B,C)
}.

/* Return true if the smallest ring is on peg A */
define smallest_on(A) = (|A|¬=0 and A[0]=0).

/* Return true if it is legal to move a ring from A to B */
define legal(A,B) = (if |A|=0 then false
                     else if |B|=0 then true
                     else A[0]<B[0]).

/* Move the topmost ring from peg A to peg B */
define move_step(A,B,C) = {
   skip and
   A <- tail(A) and
   B <- append([head(A)],B) and
   C <- C
}.

define odd(X) = (X mod 2 = 1).
```

Fig. 6.   *An iterative solution to the* Towers of Hanoi *problem.*

Finally, we can examine the *Towers of Hanoi* from a third perspective. A deeper analysis of the preceding solution reveals the full pattern of moves. Each ring (not just the smallest) follows a regular sequence of moves: it is first moved on state number 2^i, and thereafter it is moved after every 2^{i+1} steps, always in the same rotational direction. This observation leads us to conceive a concurrent solution, such as the one in Fig. 7. The program relies on parallel invocations of the procedure move$_P(i, A, B, C)$, one to generate moves for each ring *i*. On its own, each procedure move$_P$ will generate an indefinite number of moves according to the above pattern. More precisely, it will generate any prefix of an infinitely repeating pattern of moves, but

```
/* Parallel solution to the "Towers of Hanoi" problem */

define hanoi(n) = exists L,C,R : {
  len (2**n-1) and
  L=[0..n-1] and           /* list from 0 to n-1 */
  C=[] and
  R=[] and
  forall i<n : {
    if odd(n-i) then move_p(i,L,C,R) else move_p(i,L,R,C)
  } and
  always display(L,C,R)
}.


/* Generate moves for ring i, starting after 2**i-1 steps */
define move_p(i,A,B,C) = prefix {
  len (2**i-1);
  move_ring(i,A,B,C)
}.


/* Move ring from peg A to B to C to A ... Pause between moves */
define move_ring(i,A,B,C) = {
  move_step(A,B,C);
  len (2**(i+1)-1);
  move_ring(i,B,C,A)
}.


/* Move the topmost ring from peg A to peg B */
define move_step(A,B,C) = {
  skip and
  A <- tail(A) and
  B <- append([head(A)],B) and
  C <- C
}.


define odd(X) = (X mod 2 = 1).
```

Fig. 7. A concurrent solution to the Towers of Hanoi *problem.*

because the overall interval length is $2^n - 1$, each prefix must also be of this length in order to satisfy the program. We define this with the *prefix* operator.

$$\sigma \vDash \mathbf{\textit{prefix}}\ w \quad \text{iff} \quad \text{there is an interval } \sigma'$$
$$\text{such that } \sigma' \vDash w$$
$$\text{and } \sigma R_i \sigma'.$$

The prefix of a formula w is true on any initial subinterval of an interval which satisfies w.

The parallel program $p_P(n, L, C, R)$ also meets the specification:

$$\vDash p_P(n, L, C, R) \to s(L, C, R) \wedge \mathbf{\textit{len}}(2^n - 1),$$

where

$$P_P(n, L, C, R) \equiv_{\text{def}} L = \langle i{:}0 \leqslant i \langle n \rangle \wedge C = \langle\,\rangle \wedge R = \langle\,\rangle$$
$$\wedge\ \forall i \langle n{:}\textbf{\textit{if}}\ odd(n - i)\ \textbf{\textit{then}}\ move_P(n, L, C, R)$$
$$\textbf{\textit{else}}\ move_P(n, L, R, C),$$

In fact, all three solutions p_r, p_i and p_P are equivalent.

4.2 Representing Motion

How to satisfactorily model collections of independent but interacting objects in a computer program is a problem of considerable importance in such areas as process simulation and computer animation. Programming languages like *Simula* (Birtwistle *et al.*, 1973) and those based on Hewitts's 'actor' theory (1971) have met with some success, but they have little or no foundation in formal theory. Most significantly from our point of view, they have no model of time.

In this section we demonstrate a way to define moving objects in *Tempura*, but with an emphasis on method, rather than physical realism. Consider, for example, the motion of a ball as it falls to the ground. Initially the ball is stationary with its centre at point A, then the *upward* component of its velocity V decreases owing to the acceleration of gravity until finally it hits the ground. Neglecting the effects of air-resistance and so forth, such behaviour is captured in the predicate *fall*.

$$fall(A, V) \equiv_{\text{def}} \mathbf{\textit{halt}}(A_y \leqslant r) \wedge A\ \mathbf{\textit{gets}}\ \langle A_x + V_x, A_y + V_y - g/2 \rangle$$
$$\wedge\ V\ \mathbf{\textit{gets}}\ \langle V_x, V_y - g \rangle,$$

where r is the radius of the ball, and for simplicity we have restricted the motion to two dimensions. A similar predicate *rise* describes upwards motion.

$$rise(A, V) \equiv_{\text{def}} \textbf{\textit{halt}}(V_y \leqslant 0) \wedge A \textbf{\textit{ gets }} \langle A_x + V_x, A_y + V_y - g/2 \rangle$$
$$\wedge V \textbf{\textit{ gets }} \langle V_x, V_y - g \rangle.$$

The sequential composition of rising and falling specifies the trajectory of a ball which is thrown up into the air:

$$arc(A, V) \equiv_{\text{def}} rise(A, V); fall(A, V).$$

Continuing this theme, we can define a bouncing ball as the composition of several arcs:

$$bounce(A, V) \equiv_{\text{def}} \textbf{\textit{while }} V_y > 0 \textbf{\textit{ do }} \exists W : \{ W = V \wedge arc(A, W)$$
$$\wedge \textbf{\textit{fin }} (V = \langle W_x, p(W_y) \rangle) \}.$$

The vertical velocity of the ball is changed by a proportion p on each bounce. By executing this predicate in a *Tempura* program we can make a display of a bouncing ball. An output is shown in Fig. 8, where it has been assumed that the speed loss on each bounce is constant.

So far we have only considered the position of the ball and have neglected its form, but this is easily rectified. The predicate *ball* assigns a spherical shape to the ball B whose centre is at point C and whose radius is r.

$$ball(B, C, V) \equiv_{\text{def}} bounce(C, V) \wedge \Box(B\text{-}sphere(C, r)).$$

Any desired transformation might be applied in parallel with the other motion to achieve a more detailed specification. A rotational transformation would define a spinning ball, for example.

We have shown how to describe a single object. Now let us consider how to combine several objects into a larger system. Imagine a scene involving three actors: a young girl, a ball and a dog. The girl enters the scene at the start. She walks some distance with the dog following behind, then stops and throws the ball. Immediately the dog begins to run after the ball, and finally catches it. This example is rather light-hearted, but the methods we use are serious nonetheless.

Generally, an actor only appears during a certain subinterval of the scene. Thus, the definition of an actor will be of the form:

$$actor() \equiv_{\text{def}} \diamondsuit action,$$

where *action* refers to the actual motion of the actor concerned. For the actors in our scene we can be more specific.

Part of the ball's behaviour is defined above, since we assume that the ball bounces along the ground after being thrown. Thus, the predicate $b()$, which

```
/* A program to simulate rising and falling motion */
define rise(A,V) = {
  halt(V[1] <= 0) and
  A gets [A[0]+V[0],A[1]+V[1]-g/2] and
  V gets [V[0],V[1]-g]
}.

define fall(A,V) = {
  halt(A[1] <= r) and
  A gets [A[0]+V[0],max(A[1]+V[1]-g/2,0)] and
  V gets [V[0],V[1]-g]
}.

define arc(A,V) = {
  rise(A,V);fall(A,V)
}.

define bounce(A,V) = {
  while V[1]>0 do exists W : {
    W = V and
    arc(A,W) and
    fin (V=[W[0],max(-c-W[1],0)])
  }
}.
```

```
PC-Tempura (Version 2.00)

Tempura 1> load "ball".
[Reading file ball.t]
Tempura 2> run bouncing_ball().
```

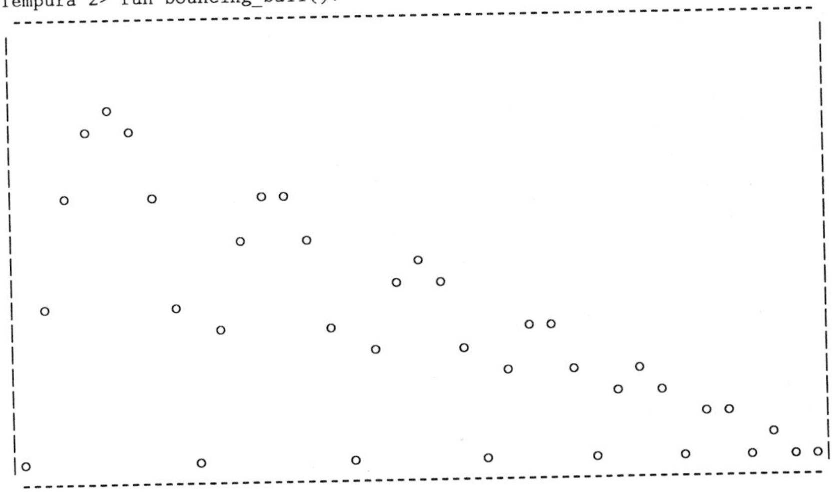

```
Done!  Computation length:  36.  Total Passes:  37.
Tempura 3>
```

Fig. 8.   Simulation of a bouncing ball in Tempura.

models the behaviour of the ball B, is defined simply as:

$$b() \equiv_{def} len(t_B); bounce(B, V); stable\ B.$$

The ball appears at time t_B, bounces for a while, and then remains static until the end of the scene. For the rest, we need only to synchronize the movement of the ball with that of the other actors.

The girl enters the scene at time t_G, and thereafter is always at position G, her movement (the variation of G) being defined by the predicate $g()$:

$$g() \equiv_{def} len(t_G); translate(G, g_0, g_1, t_B - t_G); (B = b_0 \wedge V = v_B \wedge stable\ G),$$

where the predicate $translate(G, g_0, g_1, t)$ specifies a linear translation of G from g_0 to g_1 in time t. We omit the definition. At time t_B the girl throws the ball, its initial position and velocity being defined by b_0 and v_B.

The dog spends its time in constant pursuit, first of the girl, then of the ball. Its movement is defined in terms of a predicate $follow(A, B, v)$ which specifies that the point A always moves towards the point B with a maximum velocity of v. The predicate $d()$ then gives the position D of the dog.

$$d() \equiv_{def} len\ t_D; (follow(D, G, v_D) \wedge len(t_B - t_D)); follow(D, B, v_D)$$

These definitions may be composed in parallel to form a complete scene of length s:

$$len\ s \wedge (prefix\ g()) \wedge (prefix\ b()) \wedge (prefix\ d()).$$

The **prefix** operator is used to ensure satisfiability for any value of s.

From the simplicity of these *Tempura* representations of moving objects, we may propose that the temporal logic approach compares favourably with the methods of conventional computer animation. In that area languages based on Hewitt's 'actor' theory (1971) have been very influential. Our predicates can behave in a similar way to Hewitt's actors, though in *Tempura* synchronization is achieved either implicitly (because *Tempura* programs run in lock-step) or by means of shared variables, whereas actors may only communicate by means of messages with undefined delays.

4.3 Data Transmission

With the enormous complexity of computing systems, a major concern is how to organize their descriptions so that only the necessary details are presented. The amount of detail one requires depends on one's interest, so a specification should usually admit several levels of concern. A possible solution is to organize the specification hierarchically, but it remains to find a way of relating the different levels. In this section we shall consider a small example of communication over a computer network to illustrate how the notions of *projection* and *abstraction* may be used to relate different levels of specification in *Tempura*.

We consider the problem of transmitting lines of characters from place to place, a problem which may occur, for example, in the communication of a text document. The text to be transferred (the source line) we denote by L_S, the received copy by L_R, and our aim is to repeatedly assign the contents of L_S to L_R. The following trivial predicate describes the desired behaviour:

$$level_1(L_S, L_R) \equiv_{def} L_R \textbf{ gets } L_S.$$

Suppose the transfer takes place over a wire connecting two computers where the lines are stored in buffers of 80 characters each, and suppose we transfer the text one character at a time. Then the mechanism for transmitting a single line is given by the predicate $level_2(L_S, L_R, C_S, C_R)$ which uses the variables C_S and C_R to hold the internal representations of successive characters as they are transferred to or from the line buffers. We define $level_2$ as follows:

$$level_2(L_S, L_R, C_S, C_R) \equiv_{def} s\text{-}line(L_S, C_S) \wedge C_R \textbf{ gets } C_S \wedge r\text{-}line(C_R, L_R),$$

where the send and receive predicates, *s-line* and *r-line* model the dispatch and receipt (respectively) of an 80-character line. They are defined in the corresponding *Tempura* program of Fig. 9.

The predicate $level_2$ describes a procedure for the transfer of each line. To get a complete description of the transmission at the character transfer level we need to carry out this procedure during each unit of time at the line transfer level. This is exactly what temporal projection does.

In the projected formula:

$$level_2(L_S, L_R, C_S, C_R) \textbf{ proj } level_1(L_S, L_R),$$

an interval satisfying the predicate $level_2(L_S, L_R, C_S, C_R)$ is projected onto each unit subinterval of an interval satisfying the predicate $level_1(L_S, L_R)$. In general, the projection of formula w_1 onto formula w_2 is defined as follows:

$$\sigma \vDash w_1 \textbf{ proj } w_2 \quad \text{iff} \quad \text{there are intervals } \sigma', \ldots, \sigma^{(p)}$$
$$\text{such that } \sigma = \sigma' \odot \ldots \odot \sigma^{(p)}$$
$$\text{and } \sigma^{(i)} \vDash w_1 \text{ for each } i$$
$$\text{and } \langle \sigma'_0, \sigma''_0, \ldots, \sigma^{(p)}_0, \sigma_{|\sigma|} \rangle \vDash w_2.$$

The interval composition operator $\odot$ was defined in Section 2.

Projection creates a more detailed description of a system from a higher level one. There is an 'inverse' operation known as *abstraction*, which enables the higher level specification to be recovered. Abstraction extracts subinter-

```
/************************ LINE TRANSFER LEVEL *************************/

define level1(Ls,Lr) = {
  Lr gets Ls
}.


/************************ CHARACTER TRANSFER LEVEL *******************/

define level2(Ls,Lr,Cs,Cr) = {
  r_line(Lr,Cr) and
  Cr gets Cs and                        /* unit transport delay */
  s_line(Cs,Ls)
}.

define s_line(Cs,Ls) = exists b : {     /* send a line of 80 characters */
  list(b,8) and                         /* internal send buffer b */
  b = string_to_list(Ls) and
  for i<80 do {
    skip and Cs=b[i]
  }
}.

define r_line(Lr,Cr) = exists b : {     /* receive a line of 80 characters */
  list(b,8) and                         /* internal receive buffer b */
  fin {Lr=list_to_string(b)} and
  for i<80 do {
    skip and next {b[i]=Cr}
  }
}.
```

Fig. 9.   A model of data transmission using the RS-232 protocol.

vals to give back a less detailed model. It is defined in a similar way to projection.

$$\sigma \models w_1 \; \textbf{abs} \; w_2 \quad \text{iff} \quad \text{there are intervals } \sigma', \dots, \sigma^{(p)} \text{ and } \tau$$

$$\text{such that } \tau \models w_2$$

$$\text{and } \tau = \sigma' \odot \dots \odot \sigma^{(p)}$$

$$\text{and } \sigma^{(i)} \models w_1 \text{ for each } i$$

$$\text{and } \sigma = \langle \sigma'_0, \sigma''_0, \dots, \sigma_0^{(p)}, \tau_{|\tau|} \rangle.$$

Abstraction may be used to recapture the line transfer level specification from the character transfer level.

$$\models \exists C_S, C_R : \{(\textbf{len } 80) \; \textbf{abs} \; level_2(L_S, L_R, C_S, C_R)\} \rightarrow \{\textbf{skip} \wedge level_1(L_S, L_R)\}.$$

Proceeding further down the hierarchy, we may suppose that characters are transmitted through the wire according to a collection of agreed rules.

```
/*************************** BIT TRANSFER LEVEL ****************************/

define level3(Cs,Cr,Ws,Wr) = {
  r_char(Cr,Wr) and
  Wr gets Ws and                        /* unit transport delay */
  s_char(Ws,Cs)
}.

define s_char(Ws,Cs) = exists A : {     /* send an 8 bit character */
  len 2 and clear(Ws) and               /* zero send bit */
  A <- char_to_register(Cs);            /* input char */
  for 10 times do {
    len 2 and shift(1,A,Ws)             /* shift 1 into msb of register, */
  };                                    /* lsb of register into send bit */
  always {Ws=1}                         /* return to idle */
}.

define r_char(Cr,Wr) = exists A : {     /* receive an 8 bit character */
  halt (Wr=0);                          /* wait for start bit */
  len 3 and clear_reg(A);               /* set register to zero */
  for 8 times do {
    len 2 and rotate(Wr,A)              /* shift rec'd bit into A */
  } and
  fin {Cr=register_to_char(A)}          /* output character */
}.

define rotate(W,A) = {                  /* shift W into the m.s.b. of A */
  keep_stable(A) and
  A <- append(tail(A),[W])
}.

define shift(X,A,W) = {                 /* shift the l.s.b of A into W */
  keep_stable(A) and                    /* and X into the m.s.b of A */
  keep_stable(W) and
  A <- append(tail(A),[X]) and
  W <- head(A)
}.

define clear_reg(A) = {                 /* Set register A to 0 */
  keep_stable(A) and
  A <- [0,0,0,0,0,0,0,0]
}.

define clear(A) = {                     /* Set bit value A to 0 */
  keep_stable(A) and A <- 0
}.
```

Fig. 9. Continued.

One such collection of rules is the *RS-232* protocol. According to this protocol the individual bits which make up the internal form of a character are transmitted serially along the wire at a predetermined rate. Additional control bits are used to mark the beginning and end of a character, so that there may be a pause between successive characters in the flow. Figure 10 shows the timing diagram for a single character transmission.

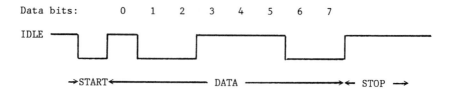

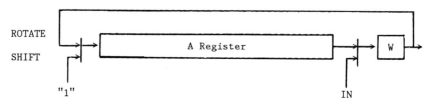

Fig. 10. Timing diagram and implementation for the RS-232 *protocol.*

The implementation of the *RS-232* protocol corresponds to a third predicate, $level_3$, where

$$level_3(C_S, C_R, W_S, W_R) \equiv_{def} s\text{-}char(C_S, W_S) \wedge W_R \textbf{ gets } W_S$$
$$\wedge \; r\text{-}char(W_R, C_R),$$

and the character transfer predicates, *s-char* and *r-char*, are defined in Fig. 9. They are both based on the hardware organization of Fig. 10, which comprises a register A and a latch W, with shift and rotate operations between them. The latches, W_S and W_R, hold the values at either end of the wire that connects the sender to the receiver. The details of the implementation are not important.

When the predicate $level_3$ is projected onto the predicate $level_2$ it gives a specification of character transfer at the bit level, and ultimately, the formula

$$level_3(C_S, C_R, W_S, W_R) \textbf{ proj } \{level_2(L_S, L_R, C_S, C_R) \textbf{ proj } level_1(L_S, L_R)\}$$

models the entire system at the bit level. By means of a similar projection of the *Tempura* programs of Fig. 9 the system can be simulated as a whole, and Fig. 11 was produced in this way (but for 8, rather than 80 character lines). Conversely, abstraction of the character transmission time from such a description gives back the higher-level predicate.

```
PC-Tempura (Version 2.00)

Tempura 1> load "rs232".
[Reading file rs232.t]
Tempura 2> run rs232_test("ABCDEFGH").

            Ls          Cs          Ws          Wr          Cr          Lr
"ABCDEFGH"              65          |
             .           .          |           |            .           .
             .           .          |           |            .           .
             .           .          |           |            .           .
             .           .          |           |            .           .
             .           .          |           |            .           .
             .           .          |           |            .           .
             .           .          |           |            .           .
             .           .          |           |            .           .
             .           .          |           |            .           .
             .           .          |           |            .           .
             .           .         ·|           |            .           .
             .           .          |           |            .           .
             .           .          |           |            .           .
             .           .          |           |            .           .
             .           .          |           |            .           .
             .           .          |           |            .           .
             .          66          |           |           65           .
             .           .          |           |            .           .
------------------------------------------------------------------------------
                                    B R E A K
------------------------------------------------------------------------------
             .           .          |           |            .           .
             .          72          |           |           71           .
             .           .          |           |            .           .
             .           .          |           |            .           .
             .           .          |           |            .           .
             .           .          |           |            .           .
             .           .          |           |            .           .
             .           .          |           |            .           .
             .           .          |           |            .           .
             .           .          |           |            .           .
             .           .          |           |            .           .
             .           .          |           |            .           .
             .           .          |           |            .           .
             .           .          |           |            .           .
             .           .          |           |            .           .
             .           .          |           |            .           .
             .           .          |           |            .           .
             .           .          |           |            .           .
             .           .          |           |           72           "ABCDEFGH"
Done!  Computation length:  176.  Total Passes:  226.
Tempura 3>
```

Fig. 11.   Simulation of the RS-232 protocol using projection.

4.4 Hardware Design

It is generally accepted that the cost of correcting faults in the design of a computer system increases rapidly as one proceeds from an initial idea to its final realization. This effect is particularly acute in the construction of large-scale integrated hardware, for once built such circuitry cannot be changed at all. Thus, time spent eliminating errors in the specification can pay dividends later on. The usual way to detect design errors is by simulation, and as a result simulation has long been an essential component of the hardware design process. In this way the designer can test out a proposed circuit by studying the behaviour of a software imitation. Obviously simulation cannot guarantee that the design is error-free, and the level of confidence it provides is decreasing all the time as the complexity of hardware increases. For the future a more formal approach is needed, as might be provided by a combination of simulation in *Tempura* and formal reasoning in ITL. This combination has been successfully applied to a variety of hardware devices, as described by Moszkowski (1983, 1985) and Hale (1985). For the present, we shall confine ourselves to consideration of a few elementary devices, starting with a trivial combinational circuit.

The output of a binary nor-gate is always equal to the negation of the disjunction of its inputs. With inputs A and B and output C such behaviour is specified by the ITL predicate *nor*:

$$nor(A, B, C) \equiv_{\text{def}} \Box(C = \neg(A \lor B)).$$

A nor-gate may be constructed from four transistors, as in Fig. 12, and the implementation captured in the predicate $nor\text{-}imp(A, B, C)$, where

$$nor\text{-}imp(A, B, C) \equiv_{\text{def}}$$
$$\exists D : \{ pt(B,1, D) \land pt(A, D, C) \land nt(A, C, 0) \land nt(B, C, 0)\},$$

and the predicates $nt(G, S, D)$ and $pt(G, S, D)$ represent n-type and p-type transistors, respectively. Each transistor behaves like a switch connecting its source (S) and drain (D) terminals. The gate terminal (G) determines whether it is switched on $(S = D)$ or off. An n-type transistor is on when $G = 1$ and off when $G = 0$; a p-type reacts in the opposite way.

It is not difficult to prove that the implementation *nor-imp* meets its specification, and this property is stated in ITL as follows:

$$\vdash nor\text{-}imp(A, B, C) \to nor(A, B, C).$$

Moreover, arbitrary transistor networks (such as *nor-imp*) can be simulated as *Tempura* programs to show their reaction to changing inputs—something which may also be done in *Prolog*, though only for a single state at a time. Unfortunately, this is not so useful in practice because delay is usually

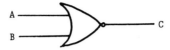

```
/* A zero-delay nor-gate */
define nor(A,B,C) = always {C = not (A or B)}.

/* A unit-delay nor-gate */
define nor1(A,B,C) = C gets not (A or B).
```

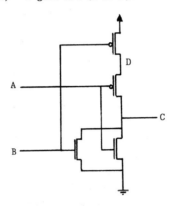

```
/* A CMOS implementation of a zero-delay nor-gate */
define nor_imp(A,B,C) = exists D : {
  pt(A,D,C) and pt(B,1,D) and nt(A,C,0) and nt(B,C,0)
}.

define pt(G,S,D) = always {if G=0 then S=D}.

define nt(G,S,D) = always {if G=1 then S=D}.
```

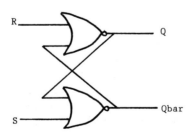

```
/* A simple latch using unit-delay nor-gates */
define latch(S,R,Q,Qbar) = {
  nor1(R,Qbar,Q) and
  nor1(S,Q,Qbar)
}.
```

Fig. 12.   The nor-gate in Tempura.

significant in the operation of real circuits. A latch, for example, cannot be conceived without delay.

We may modify our implementation of the *nor*-gate by inserting a unit-delay element before the output terminal C. It will then meet the new specification $nor_1(A, B, C)$:

$$nor_1(A, B, C) \equiv_{def} C \textbf{ gets } \neg (A \lor B).$$

From a unit-delay gate such as this, one may form a device with any throughput delay by simply projecting a different clock onto the original. For example, a nor-gate with delay n is given by:

$$nor_n(A, B, C) \equiv_{def} \textbf{prefix } \{(\textbf{len } n \land keep\text{-}stable(C)) \textbf{ proj } nor_1(A, B, C)\},$$

where

$$keep\text{-}stable(X) \equiv_{def} \exists x\text{:} \textbf{keep}(X = x).$$

The **prefix** is required because without it nor_n could only hold on an interval if its length were a multiple of n, contrary to our intention.

Two delayed nor-gates may be connected together to make an SR-latch, as depicted in Fig. 12, and the result simulated by executing the predicate *latch* as a *Tempura* program, where

$$latch(S, R, Q, Q') \equiv_{def} nor_1(R, Q', Q) \land nor_1(S, Q, Q').$$

The timing diagram of Fig. 13 was produced by such a simulation. It can be seen that the latch always responds to changes in its inputs within two time units. For example, if on an interval of length two the input S is kept high and R is kept low, the Q output will end up high.

$$\vDash \{\textbf{len } 2 \land latch(S, R, Q, Q') \land \Box(S = 1) \land \Box(R = 0)\} \rightarrow (Q \leftarrow 1).$$

Projection may also be used to create a variable delay version of the latch. For instance, a latch whose output can only change when the clock is high ($Ck = 1$) is described by the following projection:

$$\{keep\text{-}stable(Q) \land keep\text{-}stable(Q') \land \bigcirc \textbf{halt } Ck\} \textbf{ proj } latch(S, R, Q, Q').$$

In fact this is not quite correct. As in the case of the nor-gate, what we really want is the **prefix** of this formula, because we do not mean to imply that the clock should always end up high. Furthermore, for this to be a reasonable model of a physical system, certain other timing constraints must be fulfilled. Firstly, the clock must always remain high for at least two time units (to allow Q and Q' to stabilize), and secondly the inputs S and R should remain stable when the clock is high. The second constraint may be stated precisely as

```
PC-Tempura (Version 2.00)

Tempura 1> load "latch".
[Reading file latch.t]
Tempura 2> run latch_test().
State    0: Qbar   Q     R     S
State    0: |      |     |     |
State    1:    |      |  |        |
State    2: |      |     |        |
State    3: |         |  |        |
State    4: |         |  |        |
State    5: |         |  |           |
State    6: |         |  |        |
State    7: |         |  |        |
State    8: |         |  |        |
State    9: |         |  |        |
State   10: |         |  |        |
State   11: |         |     |     |
State   12: |      |        |     |
State   13:    |   |        |     |
State   14:    |   |        |     |
State   15:    |   |        |     |
State   16:    |   |     |           |
State   17: |         |  |           |
State   18: |      |     |           |
State   19: |         |  |           |
State   20: |         |  |           |
State   21: |         |  |        |
State   22: |         |  |        |
State   23: |         |  |        |
State   24: |         |  |        |
State   25: |         |  |        |

Done!  Computation length:  25.  Total Passes:  26.
Tempura 3>
```

Fig. 13. A Tempura *simulation of the latch.*

follows:

$$\boxed{a}\{(\Box(Ck = 1)) \to (\textit{stable } S \land \textit{stable } R)\}.$$

A *Tempura*-generated interval satisfying the projected latch formula and both the constraints is shown in Fig. 14.

We are not aware of another programming language which accommodates such a variety of behaviour so easily. In *Parlog* (Clark and Gregory, 1986), for example, it is possible to model a latch with the clauses of Fig. 15, but this program cannot readily be extended to handle arbitrary delays. Moreover, we must look beyond *Parlog* and first-order logic for a way to express the correctness properties of such a program.

```
PC-Tempura (Version 2.00)

Tempura 1> load "proj-latch".
[Reading file proj-latch.t]
Tempura 2> run proj_latch_test().
State    0: Qbar  Q    R    S    Ck
State    0: |     |    |    |    |
State    1: |     |    |    |    |
State    2:    |     |  |    |    |
State    3: |     |    |    |    |
State    4: |        |  |    |    |
State    5: |        |  |    |    |
State    6: |        |  |    |    |
State    7: |        |  |    |    |
State    8: |        |  |    |    |
State    9: |        |  |    |    |
State   10: |        |  |    |    |
State   11: |        |  |    |    |
State   12: |        |  |    |    |
State   13: |        |  |   |    |    |
State   14: |        |  |   |    |    |
State   15: |     |    |    |    |
State   16:    |  |    |    |    |
State   17:    |  |    |    |    |
State   18:    |  |    |   |    |    |
State   19:    |  |   |    |    |
State   20:    |  |    |    |    |
State   21: |     |    |    |    |
State   22: |        |  |    |    |
State   23: |        |  |    |    |
State   24: |        |  |    |    |
State   25: |        |  |   |    |    |
State   26: |        |  |    |    |
State   27: |        |  |    |    |
State   28: |        |  |    |    |
State   29: |        |  |    |    |
State   30: |        |  |    |    |

Done!  Computation length:  30.  Total Passes:  61.
Tempura 3>
```

Fig. 14. A Tempura *simulation of the clocked latch.*

```
/* PARLOG program to simulate an SR-latch */

mode latch(?,?,¬).
latch(s,r,q) <- nor(s,[0|q],qbar), nor(r,qbar,q).

mode nor(?,?,¬).
nor([0|x],[0|y],[1|z]) <- nor(x,y,z).
nor([0|x],[1|y],[0|z]) <- nor(x,y,z).
nor([1|x],[0|y],[0|z]) <- nor(x,y,z).
nor([1|x],[1|y],[0|z]) <- nor(x,y,z).
```

Fig. 15. An SR-latch in Parlog.

5 CONCLUSION

We have described how ITL gives rise to *Tempura*, a logic programming language which supports the idea of change, and have demonstrated its relevance to a wide range of problems. In so doing, we have shown that the 'control' part of Kowalski's equation, 'Algorithm = Logic + Control' (1979), can also be addressed in the framework of logic. *Tempura* supports algorithmic constructs such as loops, hierarchical specification and destructive assignment in an elegant and formal way, but its greatest strength is the way it handles concurrency. It is a language which offers true parallelism rather than arbitrary interleaving, and from this come the inbuilt notions of time and synchronization. Finally, *Tempura* programs are also logical specifications which will withstand formal manipulation.

Acknowledgements

This research owes a considerable debt to Ben Moszkowski for his initial work on ITL and *Tempura*. We would like to thank him for his guidance. We would also like to thank Antony Galton for carefully reading and criticizing this paper. Lastly, we acknowledge the help of the UK Science and Engineering Research Council in providing financial support.

References

Birtwistle, G. M., Dahl, O. J., Myhrhaug, B. and Nygaard, K. (1973), *SIMULA Begin* Philadelphia: Auerbach Press.
Chellas, B. F. (1980), *Modal Logic* Cambridge: Cambridge University Press.
Clark, K. L. and Gregory, S. (1986), 'Parlog: parallel programming in logic', *ACM Transactions on Programming Languages and Systems*, **8**, 1–49.
Clocksin, W. F. and Mellish, C. S. (1981), *Programming in Prolog* Berlin: Springer-Verlag.
Hale, R. W. S. (1985), 'Modelling a ring network in interval temporal logic', in *Proceedings of EuroMicro 85* Brussels, Belgium.
Hewitt, C. (1971), 'Description and theoretical analysis of PLANNER', PhD Thesis, Massachusetts Institute of Technology.
Hoare, C. A. R. (1969), 'An axiomatic basis for computer programming', *Communications of the ACM*, **12**, 576–580.
Kowalski, R. (1979), 'Algorithm = logic + control', *Communications of the ACM*, **22**, 424–436.
Moszkowski, B. C. (1983), 'Reasoning about digital circuits', PhD Thesis, Stanford University.
Moszkowski, B. C. (1985), 'A temporal logic for multi-level reasoning about hardware', *Computer*, **18**, 10–19.
Moszkowski, B. C. (1986), *Executing Temporal Logic Programs* Cambridge: Cambridge University Press.
Rescher, N. R. and Urquhart, A. (1971), *Temporal Logic* Vienna: Springer-Verlag.

4 THREE RECENT APPROACHES TO TEMPORAL REASONING

Fariba Sadri

Department of Computing,
Imperial College,
University of London

ABSTRACT

Recent work on the formalization of time and events shows evidence of a convergence of opinion amongst researchers about the desirable features of a temporal logic for artificial intelligence and database applications.

This chapter describes and discusses the following three recent approaches to temporal reasoning:

Kowalski and Sergot's calculus of events,
Lee *et al.*'s logic of time and events, and
James Allen's temporal logic.

We argue that the three approaches have much in common, and that it is not difficult to extend or specialize one to incorporate features provided by the others.

Finally, Hanks and McDermott's recent work on temporal and default reasoning is briefly discussed, with a suggestion that it merits further investigation in the light of recent results on the semantics of negation as failure.

1 INTRODUCTION

This chapter outlines and compares the three approaches of Kowalski and Sergot (1986), Lee *et al.* (1985) and Allen (1983, 1984) to the formalization of time and events.

The first and the last are formulated in first-order classical logic. The second uses the notation of Rescher and Urquhart and von Wright, but is implemented in *Prolog*.

Temporal Logics and their Applications

ISBN: 0-12-274060-2

The main intended applications of Kowalski and Sergot's calculus are updating databases and narrative understanding. Allen has investigated the use of his logic for dialogue understanding and plan-formation. Lee *et al.* have concentrated on the representation of temporal information in databases.

In our comparison of these three approaches we will focus on database applications. We will consider some of the restrictions each formalism imposes on updates that consist of event descriptions, and the information that is derivable from such transactions.

The bias in our discussion towards database applications will inevitably give an incomplete view of Allen's logic, since, unlike the others, he has concentrated on applications in plan-formation and dialogue understanding. Nevertheless, his logic can quite easily form a basis for temporal inferencing in databases. It is instructive to see how his approach can be modified to incorporate most of the characteristics offered by the other two.

In addition to these three logics, we take a brief look at Hanks and McDermott's recent work on temporal and default reasoning (Hanks and McDermott, 1985, 1986).

In Sections 2, 3 and 4 of the chapter we discuss the event calculus, Lee *et al.*'s and Allen's logics, respectively. Section 5 consists of a brief discussion on the Hanks-McDermott formalization, and in Section 6 we summarize our conclusions.

2   KOWALSKI AND SERGOT'S CALCULUS OF EVENTS

The event calculus (Kowalski and Sergot, 1986) is an approach for representing and reasoning about time and events within a logic programming framework. As the name suggests, in this calculus the notion of 'event' is central. Event descriptions imply the existence of time periods for which certain relationships hold. The holding of relationships for periods of time, in turn, implies the holding of relationships at time points.

This calculus was developed for updating databases (Kowalski, 1986) and narrative understanding as its main applications. Updates to databases consist only of additions. The effect of explicit deletion of information in conventional databases is obtained by adding new information about the end of the period of time for which the old information holds.

The following simple and informal example illustrates the general idea.

2.1   An Example

Suppose we are told that:

(1) John was employed on project P on 1/1/60.
(2) Later, on 1/2/80, he was sacked from project Q.

(3) Earlier, on 1/1/70 John was moved from project P to Q.

(4) On that same date Peter was moved from Q to P.

(5) Jill was employed on project S.

We can formulate the above information in terms of event descriptions as shown below:

We use lower-case letters for variables and function symbols, and upper-case letters for constant and predicate symbols.

(1) *E1* is the event of John being employed on P. *E1* occurs on 1/1/60. More formally:

Employ-on(*John P E1*)

Time(*E1 1/1/60*).

(2) *E2* is the event of John being sacked from Q. *E2* occurs on 1/2/80.

Sack(*John Q E2*)

Time(*E2 1/2/80*).

(3) *E3* is the event of John being moved from P to Q. *E3* occurs on 1/1/70.

Move(*John P Q E3*)

Time(*E3 1/1/70*).

(4) Similar to (3):

Move(*Peter Q P E4*)

Time(*E4 1/1/70*).

(5) Similar to (1):

Employ-on(*Jill S E5*).

Typically, events cause the start or end of periods of time. For example:

(a) An event of x being employed on project y starts a period of time for which x is assigned to y.

(b) An event of x being sacked from project y ends a period of time for which x is assigned to y.

(c) An event of x being moved from project y to z ends a period of time for which x is assigned to y, and starts a period of time for which x is assigned to z.

(a), (b) and (c) can be expressed by Horn clauses:

(a') **Assigned-to**(x y *after*(e)) **if**

Employ-on(x y e).

The term *after*(e) names the period of time started by event e. We can represent the fact that e starts the period *after*(e) by the Horn clause assertion:

Start(*after*(e) e).

(b') **Assigned-to**(x y *before*(e)) **if**

Sack(x y e).

The term *before(e)* names the period of time ended by event *e*.
The assertion

   End(*before(e)* *e*)

expresses the fact that *before(e)* is ended by *e*.

(c′) **Assigned-to**(*x y before(e)*) **if Move**(*x y z e*)
   Assigned-to(*x z after(e)*) **if Move**(*x y z e*).

Using (a′) we can conclude from sentence (1) that John is assigned to project
P for period *after(E1)*.
Pictorially:

E1   *John*   *P*

●—————————————————————

1/1/60   *after(E1)*

The end of period *after(E1)* is unspecified.

Similarly from sentences (2)–(5) it is possible to conclude the following
time periods, described pictorially:

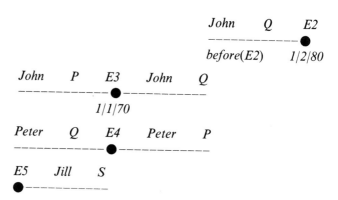

We have concluded that John is assigned to *P* for periods *after(E1)* and
before(E3), and we know of no events occurring between *E1* and *E3* that
would alter John's project assignment. In the event calculus it is possible to
conclude by default that these two periods are identical. This default
reasoning is captured by the following rule:

   The two periods of time *after(e)* and *before(e′)* are identical **if**
   x is assigned to *y* for period *after(e)* **and**
   x is assigned to *y* for period *before(e′)* **and**
   e is before *e′* **and**
   it cannot be shown that
      after(e) ends before or at the same time as
      before(e′) starts.

This kind of reasoning is non-monotonic. In the presence of new information, conclusions previously drawn by default may be automatically withdrawn. For example if we were told that John was actually moved from P and later reassigned to P in between events $E1$ and $E3$, then we could no longer infer that the periods $after(E1)$ and $before(E3)$ are identical.

The above rule can be expressed more formally as:

(d) $after(e) = before(e')$ **if**
  Assigned-to$(x\ y\ after(e))$ **and**
  Assigned-to$(x\ y\ before(e'))$ **and**
  $e < e'$ **and**
  NOT $(after(e) \ll before(e'))$,

where '$e < e'$' means 'e occurs before e'', and '$p1 \ll p2$' means 'period $p1$ ends before or at the same time as $p2$ starts', i.e.

(e) $p1 \ll p2$ **if** **End**$(p1\ e1)$ **and**
  Start$(p2\ e2)$ **and**
  $e1 \leqslant e2$.

'$e1 \leqslant e2$' expresses that $e1$ occurs before or at the same time as $e2$.

We use '$=$', '$\ll$', '$<$' and '$\leqslant$' as infixed predicate symbols for better readability.

The interpretation of the '**NOT**' in the last condition of (d) as negation by failure (Clark, 1978) gives us the default reasoning described earlier. The '**NOT**' could also be interpreted as classical negation if such default behaviour is not desirable. We could improve (d) by adding to it the extra condition that e and e' are not too far apart, and define the relation 'too far apart' appropriately.

The conclusion that $after(E1)$ is the same period as $before(E3)$ allows us to infer that

  event $E1$ starts period $before(E3)$ and
  event $E3$ ends period $after(E1)$,

using the following rules:

(f) **Start**$(before(e')\ e)$ **if** $after(e) = before(e')$
(g) **End**$(after(e)\ e')$ **if** $after(e) = before(e')$.

Notice that (d), (e), (f) and (g) give us a circular definition for the relation '$=$'. If we used rules (a'), (b'), (c'), (d)–(g) as a *Prolog* program, for example, the query '$after(E1) = before(E3)$?' would result in a non-terminating loop. This problem is avoided in the general axioms of the event calculus which we present later.

Ignoring the looping problem for now, we can similarly conclude that

after(E3) = *before(E2)*,
End(*after(E3)* *E2*), and
Start(*before(E2)* *E3*).

These conclusions give us the following picture:

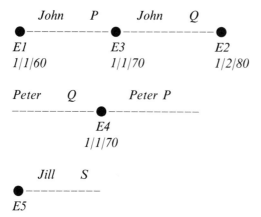

To determine if a relationship (such as 'Peter is assigned to *Q*') holds at a given instant of time we have to check if that instant occurs within a period of time for which that relationship holds. The appropriate rules for expressing this in our example are:

(h) *x* is assigned to project *y* at time *t* **if**
      x is assigned to *y* for period *p* **and**
      t is in *p*
(i) *t* is in *p* **if**
      Start(*p* *e1*) **and**
      Time(*e1* *t1*) **and**
      End(*p* *e2*)   **and**
      Time(*e2* *t2*) **and**
      t1 < *t* < *t2*.

We also need a suitable definition for the relation '<' on time points (calendar dates, in our example).

Using (h) and (i) we can conclude, for example, that

John is assigned to *P* on 1/1/65 and
      assigned to *Q* on 1/1/75.

Notice, however, that although we know Peter was moved from project *Q* to *P* on 1/1/70, we cannot conclude that he is assigned to *Q*, nor to *P*, on any date. This may seem unintuitive. The problem is that it is not known when period *after(E4)* ends, nor when *before(E4)* starts, and that therefore, using rule (i), we cannot prove that any date is in either of these two periods. On the face of it, this problem could be solved by adding to the calculus default rules such as the following:

(i*) *t* is in *p* **if**
    {**Start**(*p e*) **and**
    Time(*e t1*) **and**
    t1 < t **and**
    it cannot be shown that
    p ends before *t*} **or**
    {**End**(*p e*) **and**
    Time(*e t1*) **and**
    t < t1 **and**
    it cannot be shown that
    p starts after *t*}.

However, since in the event calculus the exact times of events and the temporal ordering of events and time points may be unknown, such a default rule can allow unacceptable consequences. The following example illustrates this point.

 Suppose that we are given the following information, represented pictorially, about four events *G1–G4*

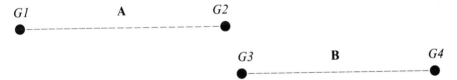

Suppose further that relationships **A**, started by *G1* and terminated by *G2*, and **B**, started by *G3* and terminated by *G4*, are incompatible. For example, **A** could be 'Margaret is Prime Minister of UK', and **B** the assertion that 'Neil is Prime Minister of UK'. Assume that the only temporal information we have about these four events is that:

 G1 occurs at time *1*,
 G4 occurs at time *5*,
 G1 occurs before *G2*,
 G3 occurs before *G4*, and
 G2 occurs before or at the same time as *G3*.

Using rule (i*), we can prove that at time *3* relationships **A** and **B** both hold, or that Margaret and Neil are both Prime Ministers of UK, which is intuitively (and constitutionally) unacceptable. This consequence is possible because we can show neither that *G2* occurs before time *3*, nor that *G3* occurs after this time. In order to avoid such problematic default conclusions we can replace (i*) by the following more stringent rules:

(j) *t* is in *p* **if**
$\qquad$**Start(*p e*) and**
$\qquad$**Time(*e t1*) and**
$\qquad$*t1* < *t* **and**
$\qquad$**NOT (End(*p e'*))**
(k) *t* is in *p* **if**
$\qquad$**End(*p e*) and**
$\qquad$**Time(*e t1*) and**
$\qquad$*t* < *t1* **and**
$\qquad$**NOT (Start(*p e'*)).**

Rules (j) and (k) allow us to conclude that Peter was assigned to project *Q* on all dates before 1/1/70, and to project *P* on all dates after 1/1/70, while avoiding the problem of rule (i*), illustrated by the example of events *G1–G4*, above.

We can add to (j) and (k) the extra condition that *t* is not too far away from *t1*. We would then need an appropriate definition for this condition.

Our project assignment example has so far illustrated a number of the general characteristics of the event calculus:

(1) Event descriptions can be assimilated in any order, independent of the order in which events actually take place. For example we can assimilate the information about John being sacked from project *Q* before that of John being moved from *P* to *Q*.

(2) Events can act as temporal references and need not be associated with absolute times. *E5*, for example, marks the start of period *after(E5)* and can act as a reference point for other periods and events, although the actual date of its occurrence is unknown.

(3) Events can be simultaneous. *E3* and *E4*, for instance, occur at the same time.

(4) Events can be partially ordered. For example, it is not known when *E5* occurs in relation to the other events.

(5) All updates are additive. The effect of deletion is obtained by adding information about the end of periods. In a conventional database, the update corresponding to event *E2* would consist of the deletion of the information 'John is assigned to *Q*'. In the event calculus the addition of the description of event *E2* allows us to infer that after *E2* John is no longer assigned to *Q*.

(6) The event calculus rules (the more general version of which we shall discuss shortly) are in Horn clause logic augmented by negation as failure. These rules can be run as a logic program in *Prolog*. However, such a program may not be very efficient and may even contain loops. Using equivalence-preserving transformations, however, it is possible to derive from these rules a more useful loop-free program.

2.2 Partial Event Descriptions

The event calculus allows events to be input with incomplete descriptions. For example, we may be told that

   (i) Peter was moved from project R or
   (ii) Peter was moved to project Q,

without being informed what project Peter was moved to in the first case or moved from in the second.

To handle such cases, events are described using binary relationships asserting only what is known and ignoring what is unknown. For example (i), above, can be represented by the assertions:

   Act($E6$ *Move*)
   Obj($E6$ *Peter*)
   Source-proj($E6$ R),

and (ii) by:

   Act($E7$ *Move*)
   obj($E7$ *Peter*)
   Destination-proj($E7$ Q),

where $E6$ and $E7$ name the events described in (i) and (ii).

This approach is based on the analysis of natural language by means of semantic networks and semantic cases.

We can modify our rules (a′), (b′), and (c′) to reflect this form of event description, and, more importantly, to draw conclusions using the minimum amount of information necessary. (c′) for example, can be modified to:

(c″) **Assigned-to**(x p *before*(e)) **if**
                    Act(e *Move*) **and**
                    Obj(e x) **and**
                    Source-proj(e p)

     Assigned-to(x p *after*(e)) **if**
                    Act(e *Move*) **and**
                    Obj(e x) **and**
                    Destination-proj(e p).

Similarly (a') and (b') can be modified:

(a″) **Assigned-to**(*x p after*(*e*)) **if**
                Act(*e Employ-on*) **and**
                Obj(*e x*) **and**
                Proj(*e p*)

(b″) **Assigned-to**(*x p before*(*e*)) **if**
                Act(*e Sack*) **and**
                Obj(*e x*) **and**
                Proj(*e p*).

We can sometimes go further and complete descriptions of partially specified events by default. For instance, suppose in our example we were given that

E6 < *E4* **and**
E4 < *E7*.

Then Peter's project assignment could be illustrated as follows:

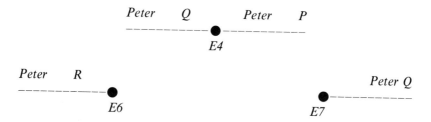

After event *E6* Peter is assigned to some unspecified project. No event is known to have affected Peter's project assignment in between events *E6* and *E4*. So in the same spirit as the default reasoning described earlier we can conclude that at event *E6* Peter was actually moved from project *R* to *Q*.

The following rule generalizes this reasoning, if the negation in its last two conditions is interpreted as failure:

(1) **Destination-proj**(*e p*) **if**
        Act(*e Move*) **and**
        Obj(*e x*) **and**
        Assigned-to(*x p before*(*e'*)) **and**
        e < *e'* **and**
        NOT (*e* < *e″* < *e'* **and**
               Assigned-to(*x p' before*(*e″*))) **and**
        NOT (*e* < *e″* < *e'* **and**
               Assigned-to(*x p' after*(*e″*))).

We can similarly conclude that the source project of *E7* is *P*, using the rule:

(m) **Source-proj**(*e p*) **if**
       Act(*e Move*) **and**
       Obj (*e x*) **and**
       Assigned-to(*x p after*(*e*′)) **and**
       e′ < *e* **and**
       NOT (*e*′ < *e*″ < *e* **and Assigned-to**(*x p*′ *before*(*e*″))) **and**
       NOT (*e*′ < *e*″ < *e* **and Assigned-to**(*x p*′ *after*(*e*″))).

As before the default conclusions can be non-monotonically withdrawn on addition of new information.

Note that default rules (l) and (m) are 'reasonable' if we assume that no-one can be assigned to more than one project at a time.

With the rules described so far, we finally obtain the following picture, after assimilating all the events specified previously concerning John's, Peter's and Jill's project assignments.

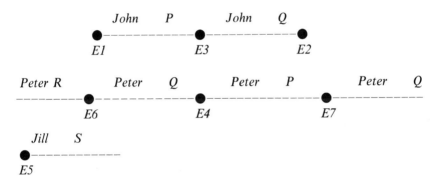

The temporal ordering of *E5* relative to the other events, and that of *E6* and *E7* relative to *E1*, *E2* and *E3*, are still unspecified.

In the next section we describe the notational conventions used in the event calculus, before presenting the general axioms.

2.3 Notation

(1) Naming time periods:
In general an event may start or end several relationships. Consider, for example, the event

'John drove his car to the airport'.

This event, call it *E*, starts two distinct time periods, for one of which John is at the airport, for the other his car is at the airport. These two periods may or may not have exactly the same duration.

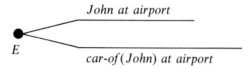

The term *after(E)* is not sufficient to name the different periods started by *E*. In the general case of the event calculus, time periods are named by terms of the form '*after(e u)*' and '*before(e u)*', where *u* is the unique relationship associated with the period.

Thus the two periods in our example are represented by the two terms:

'*after(E at(John Airport))*' and
'*after(E at(car-of(John) Airport))*'.

Here '*at*' is a function symbol, not a predicate symbol. So this formalization is in first-order logic, not higher-order logic. It can also be thought of as a weak form of first-order metalanguage where only atomic sentences (relationships) are given names.

Alternatively, it is also possible to treat an event that starts more than one relationship as a number of simultaneous events, each starting at most one single property. The same could be done for events ending more than one relationship. Event *E*, above, for example, could be thought of as two simultaneous events of John and his car getting to the airport. This approach, however, does not necessarily simplify the formalization. It would still be more convenient and more natural to input the description of a single complex event, like *E*, above. Then, in order to reason in terms of 'primitive' events, that is events that start or end at most one relationship, it would be necessary to add extra rules describing the decomposition of complex events into 'primitive' ones. We would also require additional axioms to determine the temporal relationships between 'primitive' events, given the relative ordering of the original complex events.

(2) In the general axioms of the event calculus the relation **Holds**(p) expresses that the relationship associated with time period p holds for p. Thus

$$\textbf{Holds}(after(E3\ assigned\text{-}to(\ John\ Q)))$$

expresses what we have previously written as

(*)   **Assigned-to**(*John Q after(E3)*).

It might have been more intuitive to use a two-place '**Holds**' predicate, instead, with '**Holds**($u\ p$)' expressing that relationship u holds for period p.

But that would have made the notation even more complicated. For example, statement (*), above, would have to be represented by the assertion:

Holds(*assigned-to*(*John Q*) *after*(*E3 assigned-to*(*John Q*))).

Kowalski and Sergot have opted for the one place 'Holds' predicate because of its relative simplicity, although it has a rather odd appearance.

Note that in the event calculus if 'Holds(p)' is true for some period p, and u is the relationship associated with p, then p is a *maximal* period for which u holds. More formally:

$(p1 = p2)$ **or** $(p1 \ll p2)$ **or** $(p2 \ll p1)$ **if**
                    Holds($p1$) **and**
                    Holds($p2$) **and**
                    Rel($p1$ u) **and**
                    Rel($p2$ u),

where 'Rel(p u)' expresses that u is the relationship associated with period p.

2.4 The General Axioms

Corresponding to (a″), (b″) and (c″) there are the following more general rules:

(E1) **Holds**(*before*(*e u*))   **if**   **Terminates**(*e u*)
(E2) **Holds**(*after*(*e u*))   **if**   **Initiates**(*e u*),

where 'Terminates(*e u*)' ['Initiates(*e u*)'] means event *e* terminates [initiates] relationship *u*.

In our example of project assignments we have the following definitions for 'Terminates' and 'Initiates':

Terminates(*e assigned-to*(*x y*))   **if**   Act(*e Sack*) **and**
                                              Obj(*e x*) **and**
                                              Proj(*e y*)

Terminates(*e assigned-to*(*x y*))   **if**   Act(*e Move*) **and**
                                              Obj(*e x*) **and**
                                              Source-proj(*e y*)

Initiates(*e assigned-to*(*x y*))   **if**   Act(*e Employ-on*) **and**
                                             Obj(*e x*) **and**
                                             Proj(*e y*)

Initiates(*e assigned-to*(*x y*))   **if**   Act(*e Move*) **and**
                                             Obj(*e x*) **and**
                                             Destination-proj(*e y*).

The converses of (E1) and (E2), i.e.

(E1)′ **Terminates**(*e u*)   **if**   **Holds**(*before*(*e u*)), and
(E2)′ **Initiates**(*e u*)   **if**   **Holds**(*after*(*e u*)),

are also intended to hold in the event calculus. Kowalski and Sergot do not include these in the general axioms because they are concerned with how the axioms run as a logic program, and the inclusion of (E1)′ and (E2)′ can give rise to loops in query evaluation; in any case, in practice much of the effect of these rules is obtained by using negation as failure, which uses the inverse rules implicitly rather than explicitly. Moreover, Kowalski and Sergot's description of the event calculus is biased towards database-update applications, where in general events are given as input, and one is interested in their consequences. For other applications it may be necessary to modify Kowalski and Sergot's axioms. For plan-formation, for example, where, often, the problem is to determine what events would bring about certain given properties, rules very similar to (E1)′ and (E2)′ may be needed.

The remaining general axioms are:

(E3) **Start**(*after*(*e u*) *e*)
(E4) **End**(*before*(*e u*) *e*)

Corresponding to rules (f), (g), (d) and (e) we have:

(E5) **Start**(*before*(*e u*) *e*′)   **if**   *after*(*e*′ *u*) = *before*(*e u*)
(E6) **End**(*after*(*e u*) *e*′)   **if**   *after*(*e u*) = *before*(*e*′ *u*)
(E7) *after*(*e u*) = *before*(*e*′ *u*)   **if**
              Holds(*after*(*e u*)) **and**
              Holds(*before*(*e*′ *u*)) **and**
              e < *e*′ **and**
              NOT (**Broken**(*e u e*′)),

where '**Broken**(*e u e*′)' is true if it can be shown that an event occurs between *e* and *e*′ which affects relationship *u*.

Notice that (E5) and (E6) follow from (E3) and (E4) using axioms of equality. The more general equality axioms have been avoided for the sake of efficiency.

'**Broken**' is defined by the following two rules:
(E8) **Broken**(*e u e*′)   **if**   **Holds**(*after*(*e* * *u* *)) **and**
              Related(*u u* *) **and**
              e < *e* * < *e*′
(E9) **Broken**(*e u e*′)   **if**   **Holds**(*before*(*e* * *u* *)) **and**
              Related(*u u* *) **and**
              e < *e* * < *e*′.

Two relationships are *related* if
   they are identical, or if
   they are incompatible, i.e.

(E10) **Related**($u\ u$)

(E11) **Related**($u\ u'$) **if** **Incompatible**($u\ u'$).

'Incompatible' is a domain-dependent relationship. In our example, assuming that only one person can be assigned to a specific project at any one time, we would have:

Incompatible(*assigned-to*($x\ y$) *assigned-to*($z\ y$)) **if**
    NOT ($x = z$),

together with the rule

    $x = x$

defining equality of terms.

The general rules corresponding to (h) are:

(E12) **Holds-at**($u\ t$) **if** **Holds**(*after*($e\ u$)) **and**
                          t **In** *after*($e\ u$)

(E13) **Holds-at**($u\ t$) **if** **Holds**(*before*($e\ u$)) **and**
                          t **In** *before*($e\ u$).

We use '**In**' as an infixed predicate symbol.

Rule (i) remains unchanged in the general case. Thus we have:

(E14) t **In** p **if** **Start**($p\ e1$) **and**
               Time($e1\ t1$) **and**
               End($p\ e2$) **and**
               Time($e2\ t2$) **and**
               $t1 < t < t2$.

(E15), which corresponds to rule (j), described earlier, incorporates the default assumption of forward persistence of periods:

(E15) t **In** p **if** **Start**($p\ e$) **and**
               Time($e\ t1$) **and**
               $t1 < t$ **and**
               NOT (**End**($p\ e'$)).

The symmetric default rule for backward persistence of periods, i.e.

t **In** p **if** **End**($p\ e$) **and**
              Time($e\ t1$) **and**
              $t < t1$ **and**
              NOT (**Start**($p\ e'$)),

is not included in the general axioms of the event calculus. But the rule seems to be very much in the spirit of the calculus. Because negation as failure implicitly reasons with the 'only if' half of definitions, in order for the last two rules to work correctly we require complete definitions for the '**End**' and '**Start**' predicates.The definitions, (E3)–(E6), given so far, are incomplete. In the next section we describe how the event calculus attempts to complete these definitions by providing additional rules for deriving ends and starts of periods.

2.5   Derivation of End Points For Time Periods

So far we have considered an example where the end points of periods are events that are input to the system. In some cases, however, we can infer the existence of end points without knowing of any input events corresponding to them. For example, suppose we have the following situation:

$H1$ $MP(Mary\ Labour)$

●$--------------$

$1/1/70$

$MP(Mary\ SDP)$ $H2$

$-----------------$●

$1/3/80$

i.e. Mary is a Labour Party MP (Member of Parliament) for period $after(H1)$, and an SDP Party MP for period $before(H2)$.

(For simplicity we have ignored the second parameters in $after(H1)$ and $before(H2)$.)

Suppose further that no-one can be MP for two different political parties at the same time. Then we should be able to conclude that the two periods $after(H1)$ and $before(H2)$ are disjoint, i.e. that in between January 1970 and March 1980, Mary stops being an MP for the Labour Party, and that at that same time or later, but still before event $H2$, she becomes an MP for the SDP.

Given the following definition of '**Incompatible**', rule (E16) allows us to conclude that period $after(H1)$ ends before or at the same time as $before(H2)$ starts:

Incompatible$(mp(x\ y)\ mp(x\ z))$ **if**   **NOT** $(y = z)$

(E16) $[\,fin(after(e\ u)) \leqslant init(before(e'\ u'))$ **and**
       Start$(before(e'\ u')\ init(before(e'\ u')))$ **and**
       End$(after(e\ u)\ fin(after(e\ u)))]$ **if**
                    Holds$(after(e\ u))$ **and**
                    Holds$(before(e'\ u'))$ **and**
                    Incompatible$(u\ u')$ **and**
                    $e < e'$ **and**
                    NOT (**Broken**$(e\ u\ e'))$.

In (E16) the implicit end points are named as functions of the time periods:

the term '*fin*(*after*(*e u*))' names the end of *after*(*e u*), and
'*init*(*before*(*e' u'*))' names the start of *before*(*e' u'*).

Other similar general rules describe the existence of implicit end points for time periods in a few other cases.

Note that the last condition of (E16), i.e. '**NOT**(**Broken**(*e u e'*))', is not strictly necessary. It ensures, however, that (E16) would not apply if it were possible to derive an end for *after*(*e u*) or a start for *before*(*e' u'*) by means of other rules, for example (E5) or (E6). It thus becomes unnecessary to make explicit the assumption that periods have no more than one end and one start. The explicit expression of this assumption would require equality, and would thus give rise to computational inefficiencies.

With the possibility of end points being implicit, in order for rules for completing partial event descriptions, such as (l) and (m), to work correctly, we need the following additional axioms:

Terminates(*e' u*) if **End**(*after*(*e u*) *e'*)
Initiates(*e' u*) if **Start**(*before*(*e u*) *e'*).

These rules are missing in Kowalski and Sergot's paper. Without them, (l) and (m) might incorrectly apply in the case of events with complete descriptions, and derive extra properties initiated or terminated by such events.

To set the scene for describing the logic of time and events proposed by Lee *et al.* we consider a special case for the event calculus.

2.6 A Special Case of The Event Calculus

Lee *et al.*'s logic is close to a special case of the event calculus where events are associated with times and are assimilated in the order in which they take place, the database is assumed to contain a complete record of all relevant past events, and the definition of the relation ≤ is complete. This case is similar to that assumed in conventional databases.

In this section we show how the general axioms of the event calculus can be simplified in such a case, and how from the simplified axioms, and by making explicit a number of assumptions that are implicit in the calculus, it is possible to derive rules that are very close to Lee *et al.*'s main axioms.

First, all relationships *u* are derivable in the form '**Holds**(*after*(*e u*))'. It is unnecessary to rederive them in the form '**Holds**(*before*(*e u*))'. Furthermore, we no longer need rules for deriving implicit end points.

We can thus replace (E1)–(E16) by:

(S1) **Holds**(*after*(*e u*))   **if**   **Initiates**(*e u*)

In fact, as mentioned earlier, the converse of this rule is also true, i.e. **Initiates**(*e u*) **if Holds**(*after*(*e u*)).

(S2) **Holds-at**(*u t*)   **if**   **Holds**(*after*(*e u*)) **and**
                     t **In** *after*(*e u*)

(S3) *t* **In** *after*(*e u*)   **if**   **Time** (*e t1*) **and**
                     t1 < *t* **and**
                     NOT (**End**(*after*(*e u*) *e′* **and**
                              Time(*e′ t2*) **and**
                              t2 ⩽ *t*)

(S4) **Start**(*after*(*e u*) *e*)

We also have the following rules for 'End' and 'In', but we do not need these in the discussion below:

(S5) **End**(*after*(*e u*) *e′*)   **if**   **Holds**(*after*(*e u*)) **and**
                       Terminates(*e′ u*) **and**
                       e < *e′* **and**
                       NOT (**Terminates**(*e″ u*) **and**
                             e < *e″* < *e′*)

(S6) *t* **In** *p*   **if**   **Start**(*p e1*) **and**
               Time(*e1 t1*) **and**
               End(*p e2*) **and**
               Time(*e2 t2*) **and**
               t1 < *t* < *t2*.

In addition to these simplified axioms, to derive rules similar to those of Lee *et al.*, we need the following axioms, that make explicit what is intended, but implicit, in the event calculus:

i) *t1* < *t2*   **if**   **Start**(*p e1*) **and End**(*p e2*) **and**
                  Time(*e1 t1*) **and Time**(*e2 t2*)

ii) **Terminates**(*e1 u*)   **if**   **End**(*after*(*e u*) *e1*).

Now, using (S2), (S3), (S4), (i) and (ii), we can derive the following:

(SE1) **Holds-at**(*u t*)   **if**   **Holds**(*after*(*e u*)) **and**
                         Time(*e t1*) **and**
                         t1 < *t* **and**
                         NOT (**Terminates**(*e′ u*) **and**
                              Time(*e′ t2*) **and**
                              t1 < *t2* ⩽ *t*).

Finally, from (S1) and (SE1) we obtain the following:

(SE2) **Holds-at**($u\ t$) **if** **Initiates**($e\ u$) **and**
    Time($e\ t1$) **and**
    $t1 < t$ **and**
    NOT (**Terminates**($e'\ u$) **and**
    Time($e'\ t2$) **and**
    $t1 < t2 \leqslant t$).

This rule is almost identical to one of the three main axioms of Lee *et al.*, relating events to the holding of properties at time instants.

The event calculus can be made even closer to the logic of Lee *et al.* if we assume the 'only if' direction of (SE2), i.e.

(SE3) **Holds-at**($u\ t$) $\rightarrow$
    $\exists e, t1$ [**Initiates**($e\ u$) **and**
       Time($e\ t1$) **and**
       $t1 < t$ **and**
       NOT $\exists e', t2$ (**Terminates**($e'\ u$) **and**
          Time($e'\ t2$) **and**
          $t1 < t2 \leqslant t$)],

which states that
  if a property u holds at a time instant t
  then some event, e, has started u earlier, **and**
    u has not terminated
    after the occurrence of e and before or at t.

From (SE3) and (SE2) we can derive the following rule:

(SE4) **Holds-at**($u\ t$) **if** **Holds-at**($u\ t0$) **and**
    $t0 \leqslant t$ **and**
    NOT (**Terminates**($e'\ u$) **and**
    Time($e'\ t1$) **and**
    $t0 < t1 \leqslant t$).

(SE4) is almost identical to another of Lee *et al.*'s three main axioms.

Their third main axiom, which we have not considered so far, caters for events whose exact time of occurrence is not known, but which are known to take place during given time intervals. We show how the event calculus can be extended to deal with such events, and say more about the relationship between the two logics in the course of describing that of Lee *et al.* in greater detail.

3 LEE *ET AL.*'s LOGIC OF TIME AND EVENTS

Lee *et al.* (1985) have aimed to develop a temporal system for representing and reasoning about time-dependent information and events, specifically for business database applications.

This system forms part of a larger project, called CANDID (see Lee *et al.* for a reference), which is concerned with the logical modelling of administrative databases.

Their formalization, which has been implemented in *Prolog*, is primarily in first-order classical logic, although it uses notation from Rescher and Urquhart and Von Wright (see Chapter 1 for references).

As in the event calculus, event descriptions are input to the knowledge assimilation system, and imply the holding of relationships at time instants. Events are associated with absolute or indefinite time. For example the statement

'John was hired on 10 May 1980.'

describes an event, giving its exact date, while

'Mary was fired some time during 1980.'

is an event description with an indefinite time.

To give a flavour of their approach we consider a simple example.

3.1 An Example

Suppose we have a database consisting of the following facts:

John went to Italy on 1/1/74.
He went from Italy to France on 1/1/76.
Some time during 1978 he left France and went to England.

Using Lee *et al.*'s logic we can draw the following default conclusions:

(1) John was in Italy on 1/1/75,
    since we know he went there on 1/1/74, and
    we cannot show that he left Italy any time between 1/1/74 and 1/1/75.
(2) John was in France on 29/12/77,
    by a similar argument.
(3) John was in England on 1/1/79,
    since we know that some time during 1978 he went from France to England, and we cannot show that he left England any time after the end of 1978.

However, we fail to conclude that on any date in 1978 John was in France, or that any time during that year he was in England.

The reasoning involved in this example can be informally described by the following rules:

A relationship *u* holds at time *t* **if**
some event initiates *u* at time *t0* earlier than *t* **and**
it cannot be shown that
an event terminating *u*
occurs between *t0* and *t* **and**
it cannot be shown that
an event terminating *u*
occurs during an interval *d*
such that the start of *d* is in between *t0* and *t*.

A relationship *u* holds at time *t* **if**
some event initiating *u* occurs some time during an interval *d* **and**
d ends before or at *t* **and**
it cannot be shown that
an event terminating *u*
occurs between the end of *d* and *t* **and**
it cannot be shown that
an event terminating *u*
occurs during an interval *d'*
such that the start of *d'* is in between the end of *d* and *t*.

Before describing their axioms more formally and looking at a more detailed example we summarize the notation used by Lee *et al.*

3.2 Notation

The assertion R(*t*):*u*   expresses that relationship *u* holds at time *t*. In the event calculus this is represented by the statement

$$\text{Holds-at}(u\ t).$$

In fact the event calculus form is quite close to Lee *et al.*'s Prolog implementation of R(*t*):*u*.

The term (*u1*!*u2*),   where *u1* and *u2* represent relationships, stands for the type of an event which terminates property *u1* and initiates property *u2*.
For example

$$(John\ in\ Italy!\ John\ in\ France)$$

describes an event type of John going from Italy to France.

To specify a particular event occurrence one has to associate a time with the event type. For example

$$R(date(1,1,76)):(\textit{John in Italy} ! \textit{John in France})$$

describes the particular event of John going from Italy to France on the 1st of January 1976. The notation for dates will be described shortly.

In general

$R(t):(u1!u2)$ expresses that an event of type $(u1!u2)$ occurs at time t. In the event calculus we would describe this by the following clauses:

Time$(e\ t)$
Initiates$(e\ u2)$
Terminates$(e\ u1)$,

where e names the event.

The term $(*!u)$ describes the type of an event that initiates u, and terminates **NOT** u. For example

$$(*!\textit{John in Italy})$$

describes the event type of John going to Italy.

Similarly, $(u!*)$ describes the type of an event that terminates u, and initiates **NOT** u.

The assertion

$RD(d):(u1!u2)$ states that an event of type $(u1!u2)$ occurs some time during interval d. We could describe this in the event calculus by the clauses:

Initiates$(e\ u2)$
Terminates$(e\ u1)$
During$(e\ d)$,

where e names the event, and '**During**$(e\ d)$' expresses that e occurs during interval d.

Suppose that two events of the same type, $(A!B)$, say, occur during an interval D.

In the event calculus we distinguish between the two events by giving them different names, $E1$ and $E2$, for example:

Initiates$(E1\ B)$	**Initiates**$(E2\ B)$
Terminates$(E1\ A)$	**Terminates**$(E2\ A)$
During$(E1\ D)$	**During**$(E2\ D)$.

In Lee *et al.*'s formalization to distinguish between two such events we have to associate arbitrary specific times with them:

$R(T1):(A!B)$
$R(T2):(A!B)$
During$(T1\ D)$
During$(T2\ D)$.

In the sequel we will use the term 'event' for both 'event type' and 'event occurrence', where the context makes it clear which is intended.

Dates are represented by the following terms:

$date(d, m, y)$ names the interval which is day d in month m and year y,
$date(m, y)$ names the interval which is month m in year y, and
$date(y)$ names the interval which is year y.

For example

$$R(date(1, 1, 75)):John\ in\ Italy$$

expresses that John was in Italy on 1/1/75, and

$$RD(date(1978)):(John\ in\ France!John\ in\ England)$$

expresses that some time in 1978 John left France and went to England.

Intervals of time can also be described using the function '*span*'. For example, the term

$$span(beg(date(1980)), end(date(1984)))$$

names the interval of time

from the beginning of 1980 to the end of 1984.

'*beg*' and '*end*' are function symbols.

Lee *et al.* have taken the practical approach of concentrating on calendar dates as their basic ontology of time, since in business transactions the most common temporal references are to days, months and years. Each time interval consists of infinitely many time points. The most useful time points are the ends of time intervals, for example

$beg(date(1,1,80))$ and
$end(date(1980))$.

Time instants can be related by orderings $.\leqslant.$ and $.<.$; '$t1.\leqslant.t2$' asserts that time point $t1$ precedes or equals time point $t2$, and '$t1.<.t2$' asserts that time point $t1$ precedes time point $t2$.

Time points and intervals can be related to other intervals by the relation '**During**'. '**During**$(s1\ s2)$' expresses that time $s1$ is during interval $s2$.

We are now in a position to give a formal account of the central axioms of Lee *et al.*'s formalization.

3.3   The General Axioms Describing the Holding of Relationships and the Persistence of Properties

(1) A relationship p holds at time t **if**
        p holds at an earlier time $t0$ **and**
        is not terminated in the interval $t0$ to t.

More formally:

(L1) $R(t):p$ **if** $R(t0):p$ **and**
            $t0.\leqslant.t$ **and**
            No-change$(t0\ t\ p)$.

The relation '**No-change**' is described later.

(2) A relationship p holds at time t **if**
    p is initiated at time $t0$ **and**
    $t0$ is earlier or the same as t **and**
    p is not terminated in the interval $t0$ to t.

More formally:

(L2) $R(t):p$ **if** $R(t0):(q!p)$ **and**
            $t0.\leqslant.t$ **and**
            No-change$(t0\ t\ p)$.

(3) A relationship p holds at time t **if**
        p is initiated some time during an interval d **and**
        the end of d is before or at t **and**
        p is not terminated in the interval
           from the end of d to t.

More formally:

(L3) $R(t):p$ **if** $RD(d):(q!p)$ **and**
            $end(d).\leqslant.t$ **and**
            No-change$(end(d)\ t\ p)$.

(4) '**No-change**' is described as follows:

(L4) **No-change**$(t0\ t1\ p)$ **if**
    NOT $(RD(d):(p!q)$ and $p \neq q$ and $t0.\leqslant.beg(d)$
                    and $beg(d).<.t1)$ **and**
    NOT $(R(t):(p!q)$ and $p \neq q$ and $t0.\leqslant.t$ and $t.\leqslant.t1)$.

The interpretaetion of '**NOT**' as negation by failure in (L4) gives the reasoning required for default persistence of properties.

3.4 Possible Problems With Axioms (L3) and (L4)

(1) (L3) does not take into account the possibility of property p terminating before the end of interval d. Furthermore, it is not clear how one could assimilate and use new knowledge about a more definite time of occurrence for an event which initially has only an indefinite time associated with it. To make these points clear consider the following example.

Suppose that we are given the following information:

(I1) RD($date(1980)$):($Q!P$).

Using (L3) and (L4) we can conclude that property P holds on 1/1/81, i.e.

R($date(1,1,81)$):P.

Now suppose we learn that on 27/12/80 property P terminates, i.e.

(I2) R($date(27,12,80)$):($P!*$).

Given this new information, we can still conclude that P holds on 1/1/81. This may seem unintuitive.

Other problems arise if we want to add to the knowledge base new information about the event reported in (I1). For example, suppose we learn that the event actually occurs on 1/4/80. This new information poses two problems:

(a) How do we represent it? The best we seem to be able to do is to assert:

(I3) R($date(1,4,80)$):($Q!P$).

But how do we state that (I3) and (I1) talk about the same event occurrence, rather than about two separate occurrences of the same event type?

In the event calculus the unique naming of events, and the use of binary prediates for describing events, make it easy to add new information about events specified earlier. For instance, in the present example, we can give the event reported in (I1) a unique name, H, say. Then the temporal information in (I1) may be represented by the assertion

During(H 1980),

and the new information in (I3) can be expressed by the binary relation

Time(H 1/4/80).

(b) Given (I1), (I2) and (I3), and using (L3), we can still conclude that P holds on 1/1/81. This is unacceptable.

It is not clear how (L3) could best be modified to avoid these problems. We could make the rule 'safer' by adding a couple of extra conditions to it:

(L3′) R(t):p **if** RD(d):(q!p) **and**
  end(d).≤.t **and**
  No-change(end(d) t p) **and**
  NOT (R(t1):(p!r) **and**
  p ≠ r **and**
  During(t1 d)) **and**
  NOT (RD(d1):(p!r) **and**
  p ≠ r **and**
  During(d1 d)).

We could make (L3′) even more 'robust' by adding to it the further conditions:

NOT(RD(d1):(p!r) **and** p ≠ r **and During**(d d1)) **and**
NOT(RD(d1):(p!r) **and** p ≠ r **and Overlap**(d d1)) **and** *
NOT(RD(d1):(p!r) **and** p ≠ r **and Overlap**(d1 d)),

where the relation '**Overlap**' is defined as follows:

(LE1) **Overlap**(d1 d2) **if**
  beg(d1).≤.beg(d2) **and**
  beg(d2).<.end(d1).

(Note that the condition marked (*) amongst the three extra negative conditions, above, implies and therefore makes redundant the last condition of (L3′).)

However, with or without the extra conditions, (L3′) may be too restrictive to be useful. In general the acceptability and usefulness of a default rule depends on the application and the problem domain.

(2) A feature of rule (L4) that seems unsatisfactory is that it treats two quite similar cases differently. Consider the following two situations:

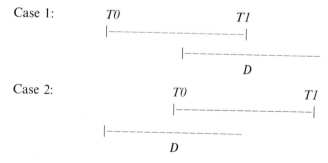

Case 1: T0   T1
  |------------------|
      |--------------------
                  D

Case 2: T0   T1
      |--------------------|
  |--------------------
          D

In both cases interval D overlaps the interval from $T0$ to $T1$. Suppose that we are given the assertions

$RD(D):(P!Q)$, and

$P \neq Q$.

Using this information and rule (L4) we can prove

'No-change($T0$ $T1$ P)'

in the second case, but not in the first.

It is easy, however, to modify (L4) so as to treat these two cases in the same way, as shown below:

(L4') **No-change**($t0$ $t1$ p) **if**
  $NOT(RD(d):(p!q)$ **and** $p \neq q$ **and**
    Overlap(d $span(t0$ $t1)$))) **and**
  $NOT(RD(d):(p!q)$ **and** $p \neq q$ **and**
    Overlap($span(t0$ $t1)$ d)) **and**
  $NOT(R(t):(p!q)$ **and** $p \neq q$ **and**
    $t0.\leqslant. t$ **and** $t.\leqslant. t1$).

To evaluate the '**Overlap**' conditions in (L4') we could use the equations

$beg(span(t$ $t')) = t$
$end(span(t$ $t')) = t'$,

and the necessary equality axioms. Alternatively we could specialize rule (LE1), as follows, for example:

Overlap($span(s$ $e)$ $span(s'$ $e')$) **if**
  $s.\leqslant. s'$ **and**
  $s'.<. e$.

3.5   The Relationship between Event Calculus Rules and Lee *et al.*'s Axioms

The event calculus rules (SE2) and (SE4), which we derived under certain assumptions, are almost 'equivalent' to (L2) and (L1), respectively. There are only two main differences:

(1) (L2) implies that if an event e initiates a property p then p holds at time of e. This is not true of (SE2). This difference could be important if we were to consider events that have duration as well as those that take place instantaneously. We will not, however, pursue this point any further.

(2) (SE2) and (SE4) do not take into account events with indefinite times. It

is not difficult to modify these rules to cater for such events. We can do so in the following way:

(SE2′) **Holds-at**(*u t*)   **if**   **Initiates**(*e u*) **and**
                                                     Time(*e t0*) **and**
                                                     t0 < *t* **and**
                                                     Not-changed(*t0 t u*)
(SE4′) **Holds-at**(*u t*)   **if**   **Holds-at**(*u t0*) **and**
                                                     t0 ≤ *t* **and**
                                                     Not-changed(*t0 t u*),

where the relation '**Not-changed**' is the event calculus version of Lee *et al.*'s '**No-change**', and is defined as follows:

Not-changed(*t0 t1 u*)   **if**
                    NOT (**Terminates**(*e u*) **and**
                             Time(*e t*) **and**
                             t0 ≤ *t* ≤ *t1*) **and**
                    NOT (**Terminates**(*e u*) **and**
                             NOT (**Time**(*e t*)) **and**
                             During(*e d*) **and**
                             Overlap(*span*(*t0 t1*) *d*)).

This rule states that
   there is no change in relationship *u* from *t0* to *t1* **if**
   it cannot be shown that
      there is an event which terminates *u* and occurs between *t0* and *t1*,
      and
   it cannot be shown that
      there is an event, with unknown definite time of occurrence, which terminates *u* and happens during an interval *d* overlapped by the interval from *t0* to *t1*.

The relation '**Overlap**' can be defined in the event calculus simply by modifying (LE1), appropriately. For example:

Overlap(*d1 d2*) **if**
                    Start-time(*d1 s1*) **and**
                    Start-time(*d2 s2*) **and**
                    End-time(*d1 e1*) **and**
                    s1 ≤ *s2* **and**
                    s2 < *e1*,

where '**End-time**(*d t*)' means that time interval *d* ends at time instant *t*, and '**Start-time**(*d t*)' expresses that interval *d* starts at time instant *t*.

We can also use a rule similar to Lee *et al.*'s (L3) in the event calculus:

(SE5) **Holds-at($u\ t$) if**
       Initiates($e\ u$) and
       NOT (Time($e\ t0$)) and
       During($e\ d$) and
       End-time($d\ t1$) and
       t1 ⩽ *t* **and**
       Not-changed($t1\ t\ u$).

Notice that, unlike (L3), rule (SE5) is not applicable in the case of an event whose definite time of occurrence is known, and therefore avoids the problem mentioned in 1(b) in the previous section.

We next look at how Lee *et al.*'s approach deals with the example of project assignments described earlier.

3.6 Application of Lee *et al.*'s Logic to the Project Assignment Example

The first four updates in the project assignment example were:

(1) John was employed on project *P* on 1/1/60.
(2) Later, on 1/2/80, he was sacked from project *Q*.
(3) On 1/1/70 John was moved from project *P* to *Q*.
(4) On that same date Peter was moved from *Q* to *P*.

There are two ways of describing these events in Lee *et al.*'s logic. For example, the event described in (1) can be represented by:

(A) R(*date(1,1,60)*):(*!assigned-to(John P)*), or by
    R(*date(1,1,60)*):*employ-on(John P)*
    together with a general rule
    R(*t*):(*!assigned-to(x y)*) **if** R(*t*):*employ-on(x y)*.

For simplicity we will use the first representation here.

    Using (A), (L2) and (L4) we can conclude that John is assigned to project *P* on 1/1/60 and on any date after that.

Pictorially:

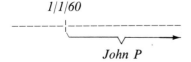

1/1/60

John P

Suppose that we assimilate the second update next. This can be represented by:

(B) R(*date(1,2,80)*):(*assigned-to(John Q)*!*).

Given (A), (B), (L2) and (L4) we can still conclude that John is assigned to *P* at any time after 1/1/60, for example on 31/1/80. This can cause problems.

Suppose that a person can be assigned to at most one project at any given time. Then the conclusion that John is assigned to *P* the day before he is sacked from *Q* is unintuitive. Suppose further that only one person can be assigned to a given project at any one time, and that we next assimilate update (4). Pictorially this would give us the following:

```
 1/1/60        John  P
  |------------------->
                Peter  P
                |------------->
              1/1/70
```

Thus we can conclude that after and on 1/1/70 both Peter and John are assigned to project *P*, a conclusion which is inconsistent with our assumption.

In the event calculus such problems are avoided by the use of rules similar to (E16), described in Section 2.4, that allow us to derive implicit ends and starts for time periods. It is not difficult to incorporate similar rules in the Lee *et al.* system. Their formalization, as it stands, however, requires events to be entered in the order they take place.

So, to be safe, we assimilate updates (2)–(4) in the order of their dates. (3) gives us:

R(*date(1,1,70)*):(*assigned-to(John P)*!*assigned-to(John Q)*).

We can now conclude, for example, that

R(*date(1,1,65)*):*assigned-to(John P)*   and
R(*date(1,1,75)*):*assigned-to(John Q)*.

But we can no longer infer that at any time on or after 1/1/70 John is assigned to *P*.

Pictorially we have the following:

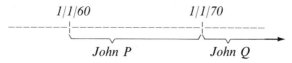

Update (4) can similarly be represented by the assertion

R(*date(1,1,70)*):
       (*assigned-to(Peter Q)*!*assigned-to(Peter P)*).

Finally, after assimilating update (2), we obtain the following picture:

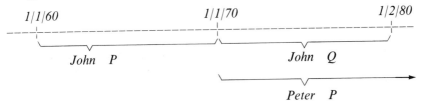

Notice that axioms (L1), (L2) and (L3) only describe persistence of relationships into the future. So we cannot conclude that at any time before 1/1/70 Peter was assigned to Q.

Notice, also, that if the date of the event reported in update (3) was not known (for example, if that event had only some arbitrary time associated with it, about which all we knew was that it was after 1/1/60), then axioms (L2) and (L4) would allow us to infer that John continues to be assigned to project P on 1/1/70 and beyond.

This example illustrates how close Lee *et al.*'s system is to the special case of the event calculus, discussed in Section 2.6, where

events are entered in the order they take place;

events have times associated with them;

properties persist forwards, only, by default; and

references to time periods over which properties hold are compiled away to give rules referring to events and the holding of properties at time instants only.

3.7 Partial Event Descriptions

Lee *et al.* do not discuss the possibility of event descriptions being incomplete. We maintain that in some cases at least they could easily deal with such events.

Suppose we are given the following partial event description:

'Mary was moved from project P on 1/1/83',

where the project Mary was moved to is not specified.

In the event calculus this event can be represented by the following assertions:

Act(E *Move*)
Obj(E *Mary*)
Source-proj(E P)
Time(E *1/1/83*),

where E is a unique name for the event.

If we later learned that Mary was in fact moved to Q, we could represent this by the assertion

Destination-proj($E\ Q$).

Using Lee *et al.*'s logic we can represent the initial information about Mary being moved from P by the following description:

$R(date(1,1,83)):move\text{-}from(Mary\ P)$.

The new information about Mary having been moved to Q can be represented by

$R(date(1,1,83)):move\text{-}to(Mary\ Q)$.

We also need the following general rules:

$R(t):(assigned\text{-}to(x\ y)!*)$ **if** $R(t):move\text{-}from(x\ y)$
$R(t):(*!assigned\text{-}to(x\ y))$ **if** $R(t):move\text{-}to(x\ y)$.

This simple approach is sufficient, in general, if we are only interested in determining whether or not a relationship holds at a given time instant.

In the event calculus event descriptions may be completed by default using the information available (see rules (l) and (m) in the description of the event calculus). We could do the same in Lee *et al.*'s logic.

For example, suppose we assume that no-one can be assigned to more than one project at a time. Then the following rule can be used to determine the destination project of a partially described '*move*' event:

(LE2) $R(t):move\text{-}to(x\ z)$ **if** $R(t):move\text{-}from(x\ y)$ **and**
                                         $R(t'):(assigned\text{-}to(x\ z)!w)$ **and**
                                         $t.<.t'$ **and**
                                         NOT$(R(t''):(q!assigned\text{-}to(x\ z))$
                                                 and $t.<.t''$ **and** $t''.<.t')$ **and**
                                         NOT$(RD(d):(q!assigned\text{-}to(x\ z))$
                                                 and $t.\leqslant.beg(d)$ **and** $beg(d).<.t')$.

Similarly, to fill in the missing information about the source project of a '*move*' event we can use the following rule:

(LE3) $R(t):move\text{-}from(x\ y)$ **if** $R(t):move\text{-}to(x\ z)$ **and**
                                         $R(t'):(w!assigned\text{-}to(x\ y))$ **and**
                                         $t'.<.t$ **and**
                                         No-change$(t'\ t\ assigned\text{-}to(x\ y))$.

Notice that if events were entered (possibly with partial descriptions) in the order they occurred then the completions of event descriptions inferred using (LE2) and (LE3) would not be default completions, that is there would never be any new information that would force the withdrawal of these completions.

To summarize, Lee *et al.* present a logic based on the concept of events and change for representing and reasoning about temporal information in business databases. In this logic events are associated with definite or indefinite times, and imply the holding of properties at time instants. Properties are assumed to persist unless it can be shown that they have terminated.

Their approach is close to the special case of the event calculus where events are associated with times and are assumed to be entered into the database in the order they take place.

Another temporal system which has much in common with both these formalisms, particularly with the event calculus, is James Allen's temporal logic, which we will now describe.

4 JAMES ALLEN'S TEMPORAL LOGIC

Like the event calculus, Allen's formalism (Allen 1983, 1984, Allen and Hayes 1984) is in first-order classical logic and is based on time intervals. Unlike the other two approaches, however, it is implemented in Lisp. He extends his temporal logic to offer an analysis of events, actions, belief, intention and causality amongst other concepts.

Allen and Koomen (1983) have investigated the application of this logic to plan-formation. The formalism has also been used in a dialogue-understanding system under development at Rochester University.

In this paper we concentrate on Allen's approach to representing and reasoning about events, time intervals and time-dependent properties. We compare this part of his logic with that of Lee *et al.* and the event calculus, and discuss how Allen's approach can be modified to incorporate some of the features offered by the latter formalism.

To introduce Allen's logic we consider the same example of project assignments we used before.

4.1 An Example

(1) John was employed on project P on 1/1/60.
(2) Later, on 1/2/80, he was sacked from project Q.
(3) On 1/1/70 John was moved from project P to Q.
(4) On that same date Peter was moved from Q to P.
(5) Jill was employed on project S.

These events can be represented in Allen's logic in the following way:

(1) The first event can be described by the assertion

$$\textbf{Occur}(employ\text{-}on^*(John\ P)\ T1)$$

(cf. R($T1$):$employ\text{-}on(John\ P)$ in Lee *et al.*'s logic.)

where '*employ-on**' is a binary function symbol. The term '*employ-on**(*John P*)' represents an event type. The assertion states that an event of John being employed on project P occurs over a time interval $T1$.

In general '**Occur**($e\ t$)' means that event e occurs over the whole of interval t. That is, there is no proper subinterval of t over which e takes place. More formally:

(A1) [**Occur**($e\ t$)   **and** $t1$ **In** t] $\rightarrow$ **NOT** (**Occur**($e\ t1$)),
    where '$t1$ **In** t' means $t1$ is a proper subinterval of t.

Notice that here, as in Lee *et al.*, to specify a particular occurrence of an event we have to specify the unique time over which it takes place.

We could, presumably, use an assertion such as

'**Time**($T1\ 1/1/60$)'

to specify the actual date of the first event in our example. Allen (1983) discusses the possibility of associating absolute time, such as calendar dates, with time intervals, and there does not seem to be any reason why this should be problematic. However as he does not pursue this very far and does not suggest any notation for representing such information we will ignore the dates of the events and only consider their ordering.

From sentences (2)–(5) we obtain the following event descriptions, where '*sack**' and '*move**' are function symbols, $T2$, $T3$, $T4$ and $T5$ name time intervals, and '$t < t'$' means that interval t is before interval t'.

(2) **Occur**($sack^*(John\ Q)\ T2$)
    $T1 < T2$
(3) **Occur**($move^*(John\ P\ Q)\ T3$)
    $T1 < T3$
    $T3 < T2$
(4) **Occur**($move^*(Peter\ Q\ P)\ T4$)
    $T3 = T4$
(5) **Occur**($employ\text{-}on^*(Jill\ S)\ T5$)

In Allen's temporal logic any two intervals are related by one of 13 mutually exclusive relations, $<$, $=$, '**Meets**', '**During**', etc. We will describe these relations later. For any two intervals t and t', we therefore know that

$$t < t' \quad \textbf{or} \quad t = t' \quad \textbf{or} \quad t \textbf{ Meets } t' \quad \textbf{or} \quad \dots$$

In our example, in particular, we can describe how interval $T5$ is related to $T1$-$T4$ by assertions of the form:

$$T5 < T1 \textbf{ or } T5 = T1 \textbf{ or } \dots$$

and

$$\vdots$$

and

$$T5 < T4 \textbf{ or } T5 = T4 \textbf{ or } \dots$$

As in the event calculus, in Allen's logic event descriptions imply the holding of relationships for time intervals. For example

> **if** an event of employing x on project y occurs over interval t **then** x is assigned to y over an interval t' such that t meets t'.

More formally:

(A2)   **Occur**($employ\text{-}on^*(x\ y)\ t)$ →
                  $\exists t'$ [t **Meets** t' **and**
                           Holds($assigned\text{-}to(x\ y)\ t')$],

where '**Holds**($u\ t$)' expresses that property u holds for interval t, and

$$\begin{array}{cc} t & t' \end{array}$$
't **Meets** t'' means $|{-}{-}{-}{-}|{-}{-}{-}{-}|$, pictorially.

Thus (1) gives us:

$$\begin{array}{ccc} T1 & John & P \end{array}$$
$$|{-}{-}{-}{-}|{-}{-}{-}{-}{-}{-}{-}{-}{-}{-}{-}{-}{-}{-}{-}{-}{-}{-}$$

Similarly:

> **if** an event of sacking x from project y occurs over a time interval t **then** there is an interval t' meeting t over which x is assigned to y, and

> **if** an event of moving x from project y to z occurs over a time interval t **then** there is an interval $t1$ meeting t over which x is assigned to y, **and** t meets an interval $t2$ over which x is assigned to z.

More formally:

(A3)   **Occur**($sack^*(x\ y)\ t)$ →
                  $\exists t'$ [t' **Meets** t **and**
                           Holds($assigned\text{-}to(x\ y)\ t')$]

(A4) **Occur**(*move**(*x y z*) *t*) →
             ∃*t1, t2*
                   [*t1* **Meets** *t* **and**
                   Holds(*assigned-to*(*x y*) *t1*) **and**
                   t **Meets** *t2* **and**
                   Holds(*assigned-to*(*x z*) *t2*)].

Thus pictorially sentences (1) to (5) give us:

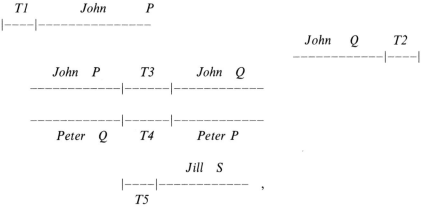

where the relative positions of *T1-T4* indicate their temporal relationships, but the exact relationship between interval *T5* and *T1-T4* is unknown.

4.2 Some Characteristics of Allen's Logic

(1) Past and future are treated symmetrically, in the sense that an event can imply information about an earlier interval which it terminates, as well as information about a later interval which it initiates. Thus, for example, we can infer that Peter was engaged on project *Q* for some time before being moved to *P*. This is also the case in the event calculus, but not in Lee *et al.*'s approach.

(2) Event descriptions, which can be viewed as updates to a database, can be entered in any order regardless of their actual order of occurrence.

(3) Unlike in the event calculus, if '**Holds**(*u t*)' is true in Allen's logic then *t* is not necessarily a maximal period for which *u* holds. In fact in Allen's formalism we have the following axiom:

(A5) **Holds**(*u t*) ↔ ∀*t'* [*t'* **In** *t* → **Holds**(*u t'*)],

which states that

> a property *u* holds for an interval *t* if and only if *u* holds for all proper subintervals of *t*.

(4) There are no axioms to allow us to conclude that the intervals met by *T1* and meeting *T3*, over which John is assigned to *P*, are equal, or even that they are either disjoint or equal.

(5) Note the difference between events and properties as illustrated by axioms (A1) and (A5). If an event *e* occurs over an interval *t*, then *e* does not occur over any proper subinterval of *t* (A1). But if a property *p* holds over *t*, then *p* holds over all subintervals of *t* (A5).

(A1) seems to prevent certain intuitive conclusions. For example, suppose that

> John wins 20 points over interval *S1*,
> loses   10 points over   *S2*, and
> wins   10 points over   *S3*,

and that

> *S1* **Meets** *S2* and
> *S2* **Meets** *S3*.

It seems reasonable to regard the event resulting from the sequential composition of these three events as an event of John winning 20 points occurring over the 'sum' of the three consecutive intervals. Pictorially:

$$
\begin{array}{ccc}
S1 & S2 & S3 \\
|{-}{-}{-}{-}{-}{-}{-}{-}| & |{-}{-}{-}{-}{-}{-}{-}{-}| & |{-}{-}{-}{-}{-}{-}{-}{-}| \\
+20 & -10 & +10
\end{array}
$$

$$
\begin{array}{c}
S \\
|{-}| \\
+20
\end{array}
$$

This composite event, however, includes the first event of John winning 20 points which occurs over interval *S1*, a proper subinterval of *S*.

Because of (A1), occurrences such as

'John was walking'

are not treated as events. If John was walking over an interval *t*, then he may also have been walking over some subintervals of *t*. Moreover, it may not be the case that he was walking over all subintervals of *t*; some time during *t* he may have paused for a rest. Thus, because of axiom (A5), it would also be problematic to represent the occurrence as a property holding over interval *t*.

To deal with such occurrences Allen introduces the idea of 'processes' into his logic (Allen, 1984). We describe these later.

In the event calculus it does not seem necessary to distinguish between occurrences that Allen calls 'events', and those he regards as 'processes'. Here, what Allen calls a process can be viewed as an event occurrence e which can be decomposed into subevents that are all different from e, but some of which are of the same type as e.

(6) In both the event calculus and in Lee *et al.*'s system properties persist by default. Allen (1983) remarks that persistence of intervals is desirable in a temporal system and speculates how this may be incorporated in his approach. However, he does not present a concrete logical formulation for such persistence.

(7) Events with partial descriptions are allowed in Allen's logic. For example we can use the assertion

$$\textbf{Occur}(move\text{-}to^*(x\ z)\ t)$$

to describe the occurrence of an event of x being moved to project z, when the project that x is being moved from is unspecified. Furthermore we can describe the relation between a partially described event and its complete version by a rule such as the following:

$$\textbf{Occur}(move\text{-}to^*(x\ z)\ t) \rightarrow$$
$$\exists y \quad \textbf{Occur}(move^*(x\ y\ z)\ t).$$

However, there are no axioms for completing partial event descriptions by default.

We now give a more detailed description of Allen's temporal logic and argue that it is easy to modify his logic to obtain characteristics similar to the event calculus.

4.3 Allen's Ontology of Time

In Allen's logic there is a single time line. Any two intervals on this time line are related by one of the 13 mutually exclusive relations described as follows:

x **Before** y $(x < y)$

x **Meets** y

x **Overlaps** y

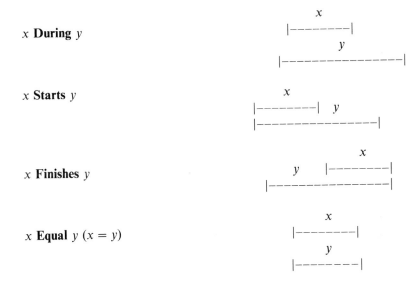

and the inverse of the first six relations. (Of course logically there is no need for the inverse relations.) All the 13 relations are definable in terms of the 'Meets' relation (Allen and Hayes, 1984).

A set of axioms describe the transitive behaviour of these temporal relations. For example:

(A6) [*t1* **Meets** *t2* **and** *t2* **During** *t3*] →
      [*t1* **Overlaps** *t3* **or**
      t1 **During** *t3* **or**
      t1 **Starts** *t3*],

and

(A7) [*t1* **Meets** *t2* **and**
      t2 **Meets** *t3*] → *t1* **Before** *t3*.

Properties hold for intervals of time. As seen earlier, the relationship 'Holds(*u t*)' expresses that property *u* holds over interval *t*. To allow properties to be represented by complex logical expressions, Allen uses function symbols '*and*', '*or*', '*not*', '*all*', and '*exists*', corresponding to the logical connectives '**and**', '**or**', '**not**', and the quantifiers ∀ and ∃.

The holding of negative properties for intervals of time is described by the following axiom:

(A8) **Holds**(*not*(*p*) *t*) ↔

      ∀*t'* [*t'* **In** *t* → **NOT** (**Holds**(*p t'*))],

where

$$t \text{ In } t' \leftrightarrow [t \text{ Starts } t' \text{ or}$$
$$t \text{ During } t' \text{ or}$$
$$t \text{ Finishes } t'].$$

For disjunctive properties we have:

(A9) **Holds**($or(p\ q)\ t$) $\leftrightarrow$

$\quad \forall t' \quad [t' \text{ In } t \rightarrow$
$\quad\quad\quad \exists s \ [s \text{ In } t' \text{ and}$
$\quad\quad\quad\quad [\textbf{Holds}(p\ s) \text{ or}$
$\quad\quad\quad\quad \textbf{Holds}(q\ s)] \ ]].$

4.4 Events Versus Processes

If an event e occurs over an interval t then there is no proper subinterval of t over which the event takes place. In Allen's logic processes differ from events in that if a process p occurs over an interval t then p must also occur over at least one proper subinterval of t. More formally:

(A10) **Occurring**($p\ t$) $\rightarrow \exists t' \quad [t' \text{ In } t \text{ and}$
$\quad\quad\quad\quad\quad\quad\quad\quad\quad \textbf{Occurring}(p\ t')],$

where '**Occurring**($p\ t$)' expresses that process p is occurring over interval t. For example the assertion '**Occurring**($falling(A)\ T$)' can be used to state that object A is falling over interval T. Notice that, as in the case of events, to identify a particular instance of a process we have to specify its time of occurrence.

Events and processes can be closely related. Suppose, for example, that '$fall(x\ y\ z)$' stands for an event of x falling from position y to z. Then the following axiom describes the relationship between a '$fall$' event and its corresponding process of '$falling$':

$$\textbf{Occur}(fall(x\ y\ z)\ t) \rightarrow \textbf{Occurring}(falling(x)\ t).$$

4.5 Modification of Allen's Logic To Incorporate Event Calculus Features

4.5.1 Default equality of intervals

To obtain an axiom similar to (E7) in the event calculus, for default equality of intervals, we can extend Allen's logic in the following way.

First we introduce a new predicate symbol '**Holds-for**' such that

Holds-for($u\ t$) is true if and only if
t is a maximal interval for which property u holds,

i.e.

(AE1) **Holds-for**(u t) ↔ [**Holds**(u t) and
              NOT ∃$t1$ [(t **In** $t1$) and
                   Holds(u $t1$)]].

As a consequence of the maximality of '**Holds-for**' we have:

[**Holds-for**(u $t1$) and
  Holds-for(u $t2$)] → [$t1 = t2$ **or**
                  $t1 < t2$ **or**
                  $t2 < t1$].

This is equivalent to:

(AE2) [**Holds-for**(u $t1$) and
     Holds-for(u $t2$) and
     NOT ($t1 < t2$) and
     NOT ($t2 < t1$)] → $t1 = t2$.

The interpretation of the '**NOT**' in the last two conditions of (AE2) as negation by failure would give us a default rule for equality of intervals.

Allen and Koomen (1983) describe a rule similar to (AE2) in English and use it in their planning algorithm, but they do not give a logical formulation of it.

In the rest of this subsection we show how (AE2) can be further transformed to resemble the event calculus rule (E7) more closely.

First, we add to the extension of Allen's logic described so far the assumption formulated in the following axiom:

(AE3) **Holds**(u t) →
              ∃t' [**Holds-for**(u t') and
              [$t = t'$ **or** t **In** t']].

(AE3) states that if a property u holds for an interval t, then either t is maximal for u or there is a maximal interval for u which contains t.

Now from (AE3) and axiom (A2) in the project assignment example, we can derive the following rule:

Occur(*employ-on**(x y) t) →
              ∃t' [**Holds-for**(*assigned-to*(x y) t') and
              [t **Meets** t' **or**
                t **Overlaps** t' **or**
                t **Starts** t' **or**
                t **During** t']].

This rule can be generalized and skolemized in the following way:

(AE4) [**Occur**(*e t*) **and Initiates**(*e u*)]
→ **Holds-for**(*u after**(*e t u*))

(AE5) [**Occur**(*e t*) **and Initiates**(*e u*)]
→ [*t* **Meets** *after**(*e t u*) **or**
t **Overlaps** *after**(*e t u*) **or**
t **Starts** *after**(*e t u*) **or**
t **During** *after**(*e t u*)],

where '*after**' is a Skolem function symbol.

(Intuitively, it may be desirable to exclude the possibility of *t* being during interval *after**(*e t u*) in (AE5).)

As with the analogous rule (E2) in the event calculus, it seems reasonable to assume the 'only if' direction of (AE4), too, that is:

(AE6) **Holds-for**(*u after**(*e t u*))
→ [**Occur**(*e t*) **and**
Initiates(*e u*)].

We in fact need this rule in transforming (AE2), as will be seen later.

In our example of project assignments, the relation '**Initiates**' is defined by the following rules:

Initiates(*move**(*x y z*) *assigned-to*(*x z*))
Initiates(*employ-on**(*x y*) *assigned-to*(*x y*)).

Similarly rules such as (A3) can be generalized, skolemized, and with the help of (AE3) transformed to the following:

(AE7) [**Occur**(*e t*) **and Terminates**(*e u*)] →
Holds-for(*u before**(*e t u*))

(AE8) [**Occur**(*e t*) **and Terminates**(*e u*)] →
[*before**(*e t u*) **Meets** *t* **or**
before*(*e t u*) **Overlaps** *t* **or**
t **Finishes** *before**(*e t u*) **or**
t **During** *before**(*e t u*)],

where '*before**' is a Skolem function symbol. (As with (AE5), earlier, we may want to exclude the possibility of *t* being during *before**(*e t u*).)

We will assume the 'only if' version of (AE7), too, that is:

(AE9) **Holds-for**(*u before**(*e t u*))
→ [**Occur**(*e t*) **and**
Terminates(*e u*)].

For the project assignment example the relation '**Terminates**' can be defined by the rules:

> **Terminates**($move^*(x \ y \ z) \ assigned\text{-}to(x \ y)$), and
> **Terminates**($sack^*(x \ y) \ assigned\text{-}to(x \ y)$).

We can now specialize rule (AE2) for default equality of intervals:

(AE10) [**Holds-for**($u \ after^*(e \ t \ u)$)   **and**
        Holds-for($u \ before^*(e' \ t' \ u)$) **and**
        $t < t'$ **and**
        NOT ($after^*(e \ t \ u) < before^*(e' \ t' \ u)$)]
                 $\rightarrow after^*(e \ t \ u) = before^*(e' \ t' \ u)$.

In deriving (AE10) from (AE2) we make use of the fact that for all e', e, t, t' and u:

> [**Holds-for**($u \ after^*(e \ t \ u)$) **and**
> **Holds-for**($u \ before^*(e' \ t' \ u)$) **and**
> $t < t'$] $\rightarrow$ **NOT** ($before^*(e' \ t' \ u) < after^*(e \ t \ u)$).

This is a consequence of (AE6), (AE5), (AE9), (AE8) and Allen's axioms describing the characteristics of the temporal relations in his logic.

Finally we can rewrite (AE10) equivalently as:

(AE11) [**Holds-for**($u \ after^*(e \ t \ u)$)   **and**
        Holds-for($u \ before^*(e' \ t' \ u)$) **and**
        $t < t'$ **and**
        NOT (**Broken**($t \ u \ t'$))]
                 $\rightarrow after^*(e \ t \ u) = before^*(e' \ t' \ u)$,

where '**Broken**' is defined as follows:

(AE12) **Broken**($t \ u \ t'$)   **if**
        Holds-for($u1 \ after^*(e1 \ t1 \ u1)$) **and**
        Related($u1 \ u$) **and**
        $t < t1 < t'$

(AE13) **Broken**($t \ u \ t'$)   **if**
        Holds-for($u1 \ before^*(e1 \ t1 \ u1)$) **and**
        Related($u1 \ u$) **and**
        $t < t1 < t'$.

The relation '**Related**' is defined exactly as in the event calculus. (Compare (AE11)–(AE13) with the event calculus rules (E7)–(E9), respectively.)

As in the event calculus, we could use equality of maximal intervals to determine if a property holds for a given period, as in the following rule, for example, which is a consequence of (AE1), (A5), (AE5), (AE6), (AE8), (AE9), and Allen's general axioms describing the transitivity characteristics of the temporal relations:

[**Holds-for**(*u after**(*e t u*)) **and**
 Holds-for(*u before**(*e' t' u*)) **and**
 *after**(*e t u*) = *before**(*e' t' u*) **and**
 t < *t''* **and**
 t'' < *t'*] → **Holds**(*u t''*).

4.5.2 Incomplete event descriptions

In this subsection we propose a way of extending Allen's logic to allow the default completion of partially described events.

Example

Let the term '*move-from**(*x y*)' stand for the event of *x* being moved from project *y*, and '*move-to**(*x z*)' for the event of *x* being moved to project *z*. Thus:

Terminates(*move-from**(*x y*) *assigned-to*(*x y*))

and

Initiates(*move-to**(*x z*) *assigned-to*(*x z*)).

Suppose that '*move-from**' and '*move-to**' provide partial descriptions of '*move**' events, where the destination project is unspecified in the former description and the source project in the latter. If we also assumed that no one is assigned to more than one project at a time, then the following additional rules could be used to complete the descriptions of these partially specified project-assignment events:

(AE14) **Occur**(*move-to**(*x z*) *t*) **if**
              Occur(*move-from**(*x y*) *t*) **and**
              Holds-for(*assigned-to*(*x z*)
                        *before**(*e t' assigned-to*(*x z*))) **and**
              t < *t'* **and**
              NOT (**Broken**(*t assigned-to*(*x z*) *t'*))

(AE15) **Occur**(*move-from**(*x y*) *t*) **if**
              Occur(*move-to**(*x z*) *t*) **and**
              Holds-for(*assigned-to*(*x y*)
                        *after**(*e t' assigned-to*(*x y*))) **and**
              t' < *t* **and**
              NOT (**Broken**(*t' assigned-to*(*x y*) *t*)).

If the negation in the conditions of (AE14) and (AE15) is interpreted as failure then these rules describe default completions of some partially specified events. For these rules to work correctly we need axioms, similar to (E16) and others in the event calculus, stating that periods of time for which incompatible relationships hold are disjoint.

In the next section we take a brief look at Hanks and McDermott's work on temporal and default reasoning.

5   HANKS AND McDERMOTT'S WORK ON TEMPORAL REASONING

Recent work by Hanks and McDermott (1985) formalizes concepts of time and default reasoning similar to those formalized in the event calculus and in the logic of Lee *et al.* These concepts include

the occurrence of events in time;
the initiation of properties by events;
the holding of properties for intervals of time; and
the persistence of properties in the absence of information to the contrary.

To formalize such default persistence of properties, Hanks and McDermott consider three different logics for default reasoning:

the nonmonotonic logic of McDermott, and McDermott and Doyle; Reiter's default logic; and
McCarthy's circumscription (see Hanks and McDermott, 1985, 1986, for references).

They conclude that all three are inherently inadequate for representing the required default reasoning, because they allow unintended models of the axioms.

As an alternative to these logics Hanks and McDermott advocate a procedural representation of the temporal axioms, implemented as a computer program (written in NISP—see Hanks and McDermott, 1985, for a reference). They argue, with the help of an intermediate inductive characterization of the program, that it computes exactly the intended model of the temporal axioms.

An alternative not considered by Hanks and McDermott is to express the temporal axioms as a logic program, that is as Horn clauses augmented by negation as failure.

Traditionally, the semantics of negation as failure has been based on Clark's completion (Clark, 1978) which, essentially, assumes that the intended program includes the 'only if' halves of all the program rules, and an appropriate equality theory. In some cases, however, such characterization allows more than one minimal model. Several recent papers (Apt *et al.*, 1986; Lifschitz, 1986; Naqvi, 1986; Przymusinski, 1986; Van Gelder, 1986) on the semantics of logic programs discuss this problem, and propose more subtle characterizations of the semantics that ensure the existence of unique minimal models, at least for a restricted class of logic programs.

It would be interesting to reconsider Hanks and McDermott's work in the light of these results on the semantics of negation as failure, and to investigate the extent to which formulating their temporal axioms as a logic program would solve the problems they have identified.

6 CONCLUSION

There are a number of similarities between the three approaches to temporal reasoning that we have looked at in detail. All three, for example, offer an analysis of the concept of events. Events are represented explicitly, can occur simultaneously, and imply the start or end of time-dependent properties. Time is also represented explicitly, rather than implicitly as it is in conventional temporal logics, where reasoning about time is performed by means of modal operators representing 'future' and 'past', for example. Furthermore all three formalisms make use of time periods, although the emphasis on periods is stronger in the event calculus and Allen's logic than in the system of Lee *et al.*

There are a number of important differences as well. For example, Allen's logic does not incorporate any default reasoning, whereas the event calculus and the logic of Lee *et al.* both allow default reasoning through the use of negation as failure. Furthermore, only in the event calculus are event occurrences given unique names. In the other two logics an event occurrence is distinguished by the type of the event and its time of occurrence. We showed through examples that giving unique names to event occurrences makes it easier to add new information about occurrences previously described, and to represent occurrences that are of the same type as one or more of their component events. Another difference is that Allen's formalism and the event calculus are more general than Lee *et al.*'s logic. The first two do not impose any restrictions on the order in which event occurrences are entered into the knowledge assimilation system, and treat past and future symmetrically. The logic of Lee *et al.*, on the other hand, requires event occurrences to be entered in the order in which they take place, and allows only for persistence of properties forwards in time.

An interesting conclusion that can be drawn from our discussion is that it is relatively easy to modify one approach in order to obtain features offered by the other two.

We attempted to show that a special case of the event calculus, where events are assimilated in the order they take place and are associated with times, is quite similar to the formalization of Lee *et al.*, ignoring differences in notation.

The logic of Lee *et al.*, in turn, can be extended to incorporate features of the event calculus. We proposed, for example, one way of dealing with events with incomplete descriptions, using the former logic. It is also easy to extend Lee *et al.*'s logic to allow the assimilation of events in any order regardless of the order in which they take place, although we did not show this.

Allen's logic and the event calculus are closely related because of their emphasis on time periods and because in both formalisms events may imply that relationships hold for periods prior to their occurrence as well as afterwards.

Allen's logic can be extended to incorporate certain characteristics of the event calculus, such as drawing default conclusions about equality of periods and the completion of partially described events.

It would be interesting to see how the event calculus might be extended to formalize some of the other concepts discussed by Allen, such as causality and intention.

Finally, a logic-programming formulation of the Hanks-McDermott temporal logic could reconcile its desirable declarative and procedural features, and facilitate its comparison with the other three approaches.

Acknowledgements

We are indebted to Robert Kowalski for many fruitful discussions, and for reading several drafts of this paper. We would also like to thank Antony Galton for his careful reading of a draft of this paper, and his many useful and constructive comments.

This work was supported by Hitachi Ltd. and the Science and Engineering Research Council.

References

In the references below *W-DD-LP* stands for the preprints of workshop on Foundations of Deductive Databases And Logic Programming, edited by Jack Minker, August 1986, Washington DC.

Allen, J. F. (1983), 'Maintaining knowledge about temporal intervals', *CACM*, **26** (11), 832–843.

Allen, J. F. (1984), 'Towards a general theory of action and time', *Artificial Intelligence*, **23**, 123–154.

Allen, J. F. and Hayes, P. J. (1984), *A Common-Sense Theory of Time* Department of

Computer Science and Philosophy, University of Rochester, Rochester, NY 14627.

Allen, J. F. and Koomen, J. A. (1983), 'Planning using a temporal world model', *IJCAI*, **2**, 741–747.

Apt, K. R., Blair, H. A. and Walker, A. (1986), 'Towards a theory of declarative knowledge', *W-DD-LP*, 546–628.

Clark, K. (1978), 'Negation as failure', in *Logic and Databases* H. Gallaire, J. Minker and J. Nicolas (eds), New York: Plenum Press.

Hanks, S. and McDermott, D. (1985), *Temporal Reasoning and Default Logics* Computer Science Research Report No. 430, Yale University.

Hanks, S. and McDermott, D. (1986), 'Default reasoning, nonmonotonic logics, and the Frame Problem', *Proceedings of the 5th National Conference on Artificial Intelligence*, AAAI, August 1986, Philadelphia, **1**, 328–333.

Kowalski, R. A. (1986), *Database Updates In The Event Calculus*, DoC 86/12, Department of Computing, Imperial College, London.

Kowalski, R. A. and Sergot, M. J. (1986), 'A logic-based calculus of events', *New Generation Computing* Ohmsha Ltd. and Springer Verlag, **4**, 67–95.

Lee, R. M., Coelho, H. and Cotta, J. C. (1985), 'Temporal inferencing on administrative databases', *Information Systems* Oxford: Pergamon Press, **10** (2), 197–206.

Lifschitz, V. (1986), 'On the declarative semantics of logic programs with negation', *W-DD-LP*, 420–433.

Naqvi, S. A. (1986), 'A logic for negation in database systems', *W-DD-LP*, 378–387.

Przymusinski, T. C. (1986), 'On the semantics of stratified deductive databases', *W-DD-LP*, 433-444.

Van Gelder, A. (1986), 'Negation as failure using tight derivatives for general logic programs', *W-DD-LP*, 712–732. Also in *Proc. 3rd IEEE Symposium on Logic Programming*, Salt Lake City, Utah.

5 THE LOGIC OF OCCURRENCE

Antony Galton

Department of Computer Science
University of Exeter

1 INTRODUCTION

A key problem for the logic of temporal discourse is how to analyse sentences which report events. In philosophy, the system introduced by Davidson (1967), in which events are represented as constituting a domain of quantification in a first-order theory, has been very influential. First-order theories of events have also been presented in AI by McDermott (1982), Allen (1984), and Kowalski and Sergot (1986). In this paper I discuss the formalization of event sentences within the modal framework exemplified by the tense logics of A. N. Prior and his followers.

The modal framework allows two kinds of tensed proposition, those prefixed by one of the weak tense-operators P and F, and those prefixed by one of the strong tense-operators H and G. If we let α stand for the proposition expressed by the sentence 'Jonathan is happy', these four tense-operators allow us to form further propositions as follows:

$$P\alpha:\text{'Jonathan has been happy'}$$
$$F\alpha:\text{'Jonathan will be happy'}$$
$$H\alpha:\text{'Jonathan has always been happy'}$$
$$G\alpha:\text{'Jonathan will always be happy'.}$$

The formal semantics of the tense-operators is a straightforward rendition in model-theoretic terms of our everyday understanding of the transformations wrought by the English tenses on the basic sentence 'Jonathan is happy'. Thus we say that 'Jonathan has been happy' is true now if 'Jonathan is happy' was true at some past time, 'Jonathan has always been happy' is true now if 'Jonathan is happy' was true at every past time, and likewise for the

Temporal Logics and their Applications

ISBN: 0-12-274060-2

future-tense sentences.[1] Formally, we posit a structure $\langle T, R, V \rangle$ consisting of a set of times T ordered by an earlier–later relation R, and a valuation function V which assigns a truth-value to each atomic proposition at each time. The meaning of the operator P is then given by the rule that $P\alpha$ is true under V at time $t \in T$ if and only if α is true under V at some time $t' \in T$ such that $t'Rt$.[2]

It is an essential part of this model that propositions are assigned truth-values at *times*; it is not essential how these times are conceived. Traditionally, they have usually been thought of as *instants*, i.e. indivisible points in time. This concept of time leads to problems when we try to analyse sentences which report events. Consider a simple example such as the event of Rosemary's eating an apple. If we are to analyse sentences reporting this event along the lines sketched above, we must posit a core proposition α to represent the simple present-tense sentence 'Rosemary eats an apple'; then 'Rosemary has eaten an apple' comes out as $P\alpha$, and 'Rosemary will eat an apple' as $F\alpha$. Applying the informal semantic analysis to these sentences, it appears that 'Rosemary has eaten an apple' must be true now if there is a past time at which 'Rosemary eats an apple' was true, and similarly for the future-tense sentence.

It is quite hard to understand just what is accomplished by this analysis. What is it for the sentence 'Rosemary eats an apple' to be true now? It cannot mean that Rosemary is eating an apple now, for the sentence which naturally expresses that (namely 'Rosemary is eating an apple') has its own past and future tenses which are distinct from the sentences we considered above. If β represents the proposition expressed by 'Rosemary is eating an apple', then $P\beta$ must correspond, straightforwardly enough, to 'Rosemary has been eating an apple', and this sentence says something different from 'Rosemary has eaten an apple' (in the latter sentence, Rosemary must have *finished* eating an apple, but in the former this is not necessary). In fact, it is not at all clear what 'Rosemary eats an apple' can possibly be understood to mean if we adhere to the notion that present-tense sentences are true or false at instants. One can *be eating* an apple at an instant, but one cannot *eat* an apple at an instant, because to eat an apple entails a succession of stages, which cannot all occur in an indivisible unit of time.

---

1. In this chapter I always translate the operator P by means of the English perfect tense ('has been', 'has eaten', etc.) rather than the preterite ('was', 'ate', etc.). In general, sentences containing the preterite presuppose some explicit or antecedently given time-reference (e.g. 'J. was happy *at such-and-such a time*'), and this cannot be formalized without first working out a theory of temporal reference. On the latter, see Section 6 below.
2. I here follow the style of modal semantics presented in Hughes and Cresswell (1984), in that no time is picked out by a model as *present*. The alternative approach, in which a model must specify a distinguished element of T as the point of primary evaluation of formulae, is illustrated by McArthur (1976).

Faced with problems like this, many workers in this area have rejected the idea that propositions should be assigned truth-values at instants, preferring instead to evaluate them on intervals (see, for example, Dowty, 1979). The sentence 'Rosemary eats an apple' now gets to be true not at individual instants, but on intervals, specifically on any interval such that the eating of an apple by Rosemary takes up just that interval and no more. On this basis, the interval semanticists have attempted to overhaul the semantics of temporal discourse in a way which avoids the difficulties alluded to above.

Although interval-based semantics has won a wide following, I believe that it rests on a confusion, the confusion between something's being true *at* a time and its being true *of* a time.[3] With instants, the confusion need not be too serious, since we can equally well say that the proposition 'Jonathan is happy' is true *at* midday, say, as that it is true *of* midday that Jonathan is happy then. The equivalence of these two ways of speaking does not carry over to intervals. We can quite well say of a given period of time *p* that Rosemary ate an apple then, meaning that her apple-eating took up exactly that period, but if we try to express this fact by saying that the sentence 'Rosemary eats an apple' is true *at* (or *on*) *p*, then we are departing radically from anything that happens in ordinary temporal discourse.[4] In ordinary discourse, we aim to express what is true *now*; we use tenses to convert facts about the past and future into facts about the present. On the interval-semanticists' picture, 'Rosemary eats an apple' never gets to be true now, since now is an instant; so the relationship between this sentence and 'Rosemary has eaten an apple' is not congruent to that between 'Jonathan is happy' and 'Jonathan has been happy'. But if a sentence cannot be true now, how can it ever be true? It seems clear to me that whatever might be meant in different contexts by the sentence 'Rosemary eats an apple', it does not play the role required of it by the interval-semanticists.

It does not seem to me, then, that interval semantics can solve the problems posed by events. What is needed is a formal representation of event-sentences which explains the logical gap created by the non-existence of a present-tensed proposition for the propositions expressed by 'Rosemary has eaten an apple' and 'Rosemary will eat an apple' to be the past and future tenses of. This gap is clearly illustrated in Table I. The 'states' column shows the straightforward case of a sentence reporting a state of affairs; the 'events' columns show how an event can be referred to in two ways, either as an event

---

3. Cf. Tichý (1985), p. 269: 'Instead of construing [predicates] as signifying properties which individuals instantiate *at* intervals, they can be construed as *relations* between individuals and intervals' (first italics mine). Tichý is here advancing an argument broadly similar to mine.

4. In fact, the commonest way in which we ascribe events to periods is different from either of these: when we say that Rosemary ate an apple yesterday, we do not mean that she took all day to eat the apple, only that she ate it at some time during the day.

in progress, or as a completed whole. In the former case, the event is treated as a state, which may accordingly be located, like any other state, in the past, present or future; in the later case, however, location in the present is not possible, so only past and future sentences are given.

Table I

| | STATES | EVENTS | |
		IN PROGRESS	COMPLETE
PAST	*Jonathan has been happy*	*Rosemary has been eating an apple*	*Rosemary has eaten an apple*
PRESENT	*Jonathan is happy*	*Rosemary is eating an apple*	—
FUTURE	*Jonathan will be happy*	*Rosemary will be eating an apple*	*Rosemary will eat an apple*

What immediately strikes one about Table I is that events give rise to a richer set of semantic structures than states. For the state of Jonathan's being happy, we have just three propositions in the table; for the event of Rosemary's eating an apple, we have five. Of the three propositions relating to Jonathan's being happy, the present-tense one is clearly basic in the sense that the others can be straightforwardly derived from it by the addition of appropriate qualifications. Of the five propositions relating to Rosemary's eating an apple, however, none can be regarded as basic in this sense. On the other hand, all of them clearly do contain a common core of meaning, namely the idea of Rosemary's eating an apple. The conclusion to be drawn is that this meaning is not something that can be expressed as the entire content of some proposition. Rather, we must invoke some other kind of formal entity to be its bearer.

In *The Logic of Aspect* (Galton, 1984), I introduced the category of *event-radicals* to fulfil this role. An event-radical is a complete expression (in Frege's sense) that is neither a proposition nor a proper name. It 'denotes' an event. Thus the event-radical 'Rosemary-EAT-an-apple' denotes the event of Rosemary's eating an apple. 'Event' is here to be understood to refer to an event-*type*, i.e. a generic occurrence which may occur any number of times, or not at all. It is not an event-*token*, which is a specific occurrence, in a particular place at a particular time, which in the nature of things it makes no sense to speak of as occurring more than—or less than—once. I shall refer to event-tokens as *occurrences*. An example of an occurrence is the occasion of Rosemary's eating an apple on the afternoon of her second birthday. This

THE LOGIC OF OCCURRENCE

occurrence is an occurrence *of* the event-type denoted by the radical 'Rosemary-EAT-an-apple'.[5]

Having introduced event-radicals, we must next introduce formal apparatus with which to construct sentences which contain them. This apparatus consists of a set of *aspect operators* which can be applied to event-radicals to yield propositions. Three operators are required, which convert the radical *Rosemary-EAT-an-apple* into the three sentences *Rosemary has eaten an apple*, *Rosemary is eating an apple*, and *Rosemary will eat an apple*. The other two sentences involving this radical in Table I are obtained by applying the ordinary past and future tense-operators to *Rosemary is eating an apple*. The three aspect operators are called the *perfect, progressive* and *prospective* operators respectively. I denote them by the symbols *Perf, Prog*, and *Pros*. Writing α for 'Jonathan is happy', and E for the radical 'Rosemary-EAT-an-apple', the sentences in Table I are formalized as in Table II.

Table II

	STATES	EVENTS	
		IN PROGRESS	COMPLETE
PAST	$P\ \alpha$	$P\ Prog\ E$	$Perf\ E$
PRESENT	α	$Prog\ E$	—
FUTURE	$F\ \alpha$	$F\ Prog\ E$	$Pros\ E$

So far, I have only discussed the syntactic requirements of the new theory. In addition, we must consider the semantics. Informally, the meanings of *Perf* and *Pros* are easy enough to state, namely that *PerfE* is true now if some occurrence of the event denoted by E is wholly in the past, and *ProsE* is true now if some occurrence of that event is wholly in the future. *Prog* presents more difficulty, because there are many different ways in which an event may be said to be in progress. The simplest interpretation is that *ProgE* is true now if some occurrence of E is partly in the past, partly present, and partly future. Under this interpretation, to say that Rosemary is eating an apple is to say that she has begun to eat an apple, and will finish eating it. I call this the *broad closed* sense of the progressive. Very often, though, we use the progressive without wishing to assert the second clause of this definition. Rather, what we

---

5. It is important, incidentally, to distinguish an event which has only one occurrence from that occurrence itself (cf. *The Logic of Aspect*, p. 56). Some events by their nature cannot occur more than once, e.g. the event of someone's being born, or dying, or indeed doing anything for the first time, for the second time, etc. Such an event may fittingly be called *once-only* event; and it is not to be identified with its unique occurrence.

mean by 'Rosemary is eating an apple' is that Rosemary has begun to eat an apple, and there are grounds for believing that she will finish eating it. The nature of these grounds varies from case to case. It may be that Rosemary *intends* to eat the whole apple, or that she has been observed to be in the *habit* of finishing any apple she starts, and so on. The exact statement of these different kinds of ground is very tricky, involving as it does notions like intention, habit, etc., which experience shows are logically rather intractable. In *The Logic of Aspect* I proposed that the first interpretation I have mentioned here, the broad closed sense, is fundamental, and that the other senses can be derived from it by the addition of various kinds of modal operators. I see no reason to change this view. In what follows, then, I shall always understand *Prog* to express the broad closed sense of the progressive.

Having established the informal requirements for the semantical account of the aspect operators, we now turn to their formalization. Given an underlying semantic model of the kind standardly used for tense logic (i.e. a structure of the kind $\langle T, R, V \rangle$ alluded to earlier), how are event-radicals to be interpreted in this model?

In order to see the kind of thing that is required, consider again the way in which propositions are interpreted in tense logic. No attempt is made to analyse the *meaning* of a basic proposition like 'Jonathan is happy'. This proposition functions as an unanalysable unit from the point of view of tense logic, any account of how its meaning is derived from the meanings of its constituent parts lying outside the scope of that logic. What tense logic concerns itself with is rather the *temporal incidence* of the proposition, i.e. the times at which it is true. From the point of view of tense logic, this is the only feature of a proposition that matters, and propositions with the same temporal incidence in a given model count as equivalent in that model.

This kind of simplification is typical of all enterprises that go under the name of logic. In the propositional calculus, for example, propositions are not analysed beyond their truth-functional structure, and that structure is blind to everything except the truth-values of its constituents. So the only feature of a proposition which really matters in propositional logic is its truth-value. Similarly, in the predicate calculus, what matters for the semantic definition of a predicate is the set of individuals falling under it, i.e. its extension, and predicates with the same extension in a model count as equivalent in that model.

Applying the same considerations to event-radicals, it seems pretty clear that what we ought to consider is *their* temporal incidence, i.e. an event is to be represented in a model in terms of the times at which it happens. The exact manner in which this is accomplished in my system is detailed below. It is in broad agreement with the schemes proposed by McDermott (1982) and Allen (1984); but these authors, especially the former, seem to be in some confusion as to the exact nature of the simplification thereby effected. (Cf. my remarks above pp. 22–23). Their systems also differ from mine in being based on predicate calculus rather than on tense logic.

To be precise, what is presented in this chapter is a *logic of occurrence*, i.e. the logic of the operators *Perf*, *Prog*, and *Pros* which serve to express the occurrence of events in time. It is not a *logic of events*, if by that is meant a scheme for analysing the events into their constituent states.[6] In the final section of this paper, I made some remarks about how the logic of occurrence might be extended to incorporate such analyses.

The organization of the rest of this chapter is therefore as follows. In Section 2, I define the language of Event Logic and specify formal model theories for several different versions of it. In Section 3, I present corresponding proof theories. In Sections 4 and 5, I give soundness and completeness proofs for these proof theories. In Section 6, I discuss how the formalism might be adapted to encompass explicit temporal reference, and finally, in Section 7, I consider the further analysis of events themselves.

2 THE LANGUAGE OF EVENT LOGIC AND ITS FORMAL SEMANTICS

The first stage in the process of formalization is to specify the language in which event-logical formulae are to be constructed. This consists of a lexicon, rules of formation, and definitions.

The Lexicon

This consists of the following primitive symbols:

Connectives: $\neg$, $\wedge$.
Tense Operators: P, F.
Aspect Operators: *Perf*, *Prog*, *Pros*.
Event Radicals: E_1, E_2, E_3, ...

The Rules of Formation

The well-formed formulae (*wffs*) of Event Language are defined by the rules:

(I) If E is an event-radical then *PerfE*, *ProgE*, and *ProsE* are all wffs (*atomic* wffs);

(II) If α and β are wffs then so are $\neg \alpha$, $(\alpha \wedge \beta)$, $F\alpha$ and $P\alpha$.

---

6. Despite this, I retain the name 'event logic' introduced in *The Logic of Aspect*. 'Aspect logic' might be a better name, suggesting as it does a relationship with the linguistic category of aspect comparable to the admittedly rather indirect relationship between tense logic and the linguistic category of tense. In either case, what is provided by the logic requires a good deal of extra-logical manipulation to be made suitable as a tool for linguistic analysis.

Definitions

Further symbols are defined, in standard fashion, as follows:

$$(\alpha \vee \beta) =_{df} \neg(\neg\alpha \wedge \neg\beta)$$
$$(\alpha \rightarrow \beta) =_{df} \neg\alpha \vee \beta$$
$$(\alpha \equiv \beta) =_{df} (\alpha \rightarrow \beta) \wedge (\beta \rightarrow \alpha)$$
$$G\alpha =_{df} \neg F \neg \alpha$$
$$H\alpha =_{df} \neg P \neg \alpha$$

The intended meanings of wffs of Event Language are formally specified by means of the following model theory. First, we assume that time is linear. We begin with the notion of a *Linear Temporal Frame*, which is a structure $\langle T, < \rangle$, where T is any set (the elements of T being the 'times' of the model), and $<$ is a total ordering on T, i.e. a relation with the following properties:

Irreflexivity
For no t in T does $t < t$ hold.

Transitivity
For every t, t', t'' in T, if $t < t'$ and $t' < t''$ then $t < t''$.

Linearity
For every t, t' in T, either $t < t'$, $t = t'$, or $t' < t$.

The symbol '$<$' may be read 'precedes' or 'is earlier than'.

Next, we identify what elements of the structure are to correspond to events. Since, as we have remarked, we do not expect our treatment to extend beyond the facts of temporal incidence of events, it will be enough to identify each occurrence by the time at which it occurs. At this point, for reasons that will emerge, I take the somewhat unusual step of representing an occurrence by an ordered pair $\langle B, A \rangle$, where B is to be understood as the set of times *before* the occurrence, and A the set of times *after* it. The formal definition is:

A *formal occurrence* in $\langle T, < \rangle$ is a pair $\langle B, A \rangle$, where B and A are non-empty subsets of T such that:

(FO1) Every element of B precedes every element of A.
(FO2) Every element of T which precedes an element of B is itself an element of B.
(FO3) Every element of T which is preceded by an element of A is itself an element of A.

The net effect of this definition is that B is an initial segment of T under the precedence relation, while A is a final segment; and B and A are disjoint.

Note that the definition of a formal occurrence leaves open the possibility that B and A jointly exhaust T, i.e. that $B \cup A = T$. In such a case we would have a formal occurrence with the property that at any time it either has occurred already or is yet to occur. Such an occurrence is at no time in the process of occurring; rather, it marks the boundary between two states of affairs, e.g. a body's starting to move marks the boundary between a stretch of time throughout which it is at rest and a stretch of time throughout which it is in motion, there being no third possible state to mediate between these two. This kind of occurrence takes no time, and being therefore point-like may fittingly be designated *punctual*. An occurrence which is not punctual has duration, and may accordingly be called *durative*. The definition of a formal occurrence has been chosen specifically to facilitate the representation of punctual occurrences.

An event may now be represented by the collection of all its occurrences, so we define a *formal event* in $\langle T, < \rangle$ to be nothing other than a set of formal occurrences in $\langle T, < \rangle$. There is no requirement that this set should be non-empty; and indeed the empty formal event must correspond to an event which has no occurrences, just as in the standard model theory for first-order logic the empty set corresponds to a predicate which is not true of anything. The occurrences of a formal event may be all punctual, all durative, or a mixture of both. We may speak of the formal event itself in these cases as punctual, durative, or mixed. Mixed formal events are of little interest, however, since we do not normally classify events in the world in a way which corresponds to them.

An *event-logical model* can now be defined as a triple $\langle T, <, I \rangle$, where $\langle T, < \rangle$ is a linear temporal frame and I is a mapping from the set of event radicals of the language into the set of formal events in $\langle T, < \rangle$. We shall usually denote such models as M, M', M'', etc.; where necessary, we distinguish the T, $<$, and I pertaining to a particular model M as T_M, $<_M$, and I_M.

The truth-definition for event language can now be stated as follows:

Atomic wffs

$\quad$ M $\vDash$ *PerfE*$[t]$ iff $t \in A$ for some $\langle B, A \rangle \in I_M(E)$.

$\quad$ M $\vDash$ *ProsE*$[t]$ iff $t \in B$ for some $\langle B, A \rangle \in I_M(E)$.

$\quad$ M $\vDash$ *ProgE*$[t]$ iff $t \in c(B) \cap c(A)$ for some $\langle B, A \rangle \in I_M(E)$.

(Here '$c(X)$' denotes the set-theoretic complement of X).

Compound wffs

$M \vDash \neg\alpha[t]$ iff it is not the case that $M \vDash \alpha[t]$.

$M \vDash \alpha \wedge \beta[t]$ iff $M \vDash \alpha[t]$ and $M \vDash \beta[t]$.

$M \vDash P\alpha[t]$ iff $M \vDash \alpha[t']$ for some $t' \in T_M$ such that $t' <_M t$.

$M \vDash F\alpha[t]$ iff $M \vDash \alpha[t']$ for some $t' \in T_M$ such that $t <_M t'$.

From the definitions of $\vee$, $\rightarrow$, $\equiv$, H, and G we can deduce that:

$M \vDash \alpha \vee \beta[t]$ iff either $M \vDash \alpha[t]$ or $M \vDash \beta[t]$.

$M \vDash \alpha \rightarrow \beta[t]$ iff $M \vDash \beta[t]$ unless $M \vDash \neg\alpha[t]$.

$M \vDash \alpha \equiv \beta[t]$ iff '$M \vDash \alpha[t]$' and '$M \vDash \beta[t]$' have the same truth-value.

$M \vDash H\alpha[t]$ iff $M \vDash \alpha[t']$ for every $t' \in T_M$ such that $t' <_M t$.

$M \vDash G\alpha[t]$ iff $M \vDash \alpha[t']$ for every $t' \in T_M$ such that $t <_M t'$.

Note also that from the truth-definition for atomic wffs it is a simple deduction that

$$\{t \in T_M | M \vDash PerfE[t]\} = \bigcup\{A | \langle B, A \rangle \in I_M(E)\}.$$

$$\{t \in T_M | M \vDash ProsE[t]\} = \bigcup\{B | \langle B, A \rangle \in I_M(E)\}.$$

$$\{t \in T_M | M \vDash ProgE[t]\} = \bigcup\{c(B) \cap c(A) | \langle B, A \rangle \in I_M(E)\}.$$

We may describe a model M as *punctual* (*durative*) if $I_M(E)$ is punctual (durative) for every E. It is *once-only* if $I_M(E)$ contains at most one member for every E.

We shall use the symbol $\mathcal{M}$ with different subscripts to denote *classes* of models. In particular we shall use:

$\mathcal{M}_E$ to denote the class of *all* event-logical models;
$\mathcal{M}_{EP}$ to denote the class of all *punctual* event-logical models;
$\mathcal{M}_{EO}$ to denote the class of all *once-only* event-logical models;
$\mathcal{M}_{ED}$ to denote the class of all *durative* event-logical models.

Obviously, $\mathcal{M}_{EP}$, $\mathcal{M}_{EO}$, and $\mathcal{M}_{ED}$ are all subsets of $\mathcal{M}_E$.

For any class of models $\mathcal{M}$ we write $\vDash_{\mathcal{M}}\alpha$ to mean that $M \vDash \alpha[t]$ for every $M \in \mathcal{M}$ and $t \in T_M$. If Σ is a set of wffs we write $\Sigma \vDash_{\mathcal{M}}\alpha$ to mean that $M \vDash \alpha[t]$ for every $M \in \mathcal{M}$ and $t \in T_M$ such that $M \vDash \sigma[t]$ for every $\sigma \in \Sigma$.

If we ignore the internal structure of the atomic wffs, event language becomes the ordinary language of tense logic. We can in this way regard a model for event language as giving rise to a model for tense language. Such a model is, in general, a triple $\langle T, <, V \rangle$, where V is a valuation function which assigns a truth-value (say 0 or 1) to each of the atomic wffs of the language at each time in T. The truth-definition for a model of this kind is given for atomic wffs by means of the condition $M \vDash \alpha[t]$ iff $V(\alpha, t) = 1$, and for the compound wffs by the same set of rules as was given for the compound wffs of event language above. We can say that an event-logical model $M = \langle T, <, I \rangle$ *generates* the tense-logical model $gen(M) = \langle T, <, V_I \rangle$ by way of the following identifications:

$V_I(PerfE, t) = 1$ iff $t \in A$ for some $\langle B, A \rangle \in I(E)$.
$V_I(ProsE, t) = 1$ iff $t \in B$ for some $\langle B, A \rangle \in I(E)$.
$V_I(ProgE, t) = 1$ iff $t \in c(B) \cap c(A)$ for some $\langle B, A \rangle \in I(E)$.

It is a straightforward induction proof from these stipulations to the result that $gen(M) \vDash \alpha[t]$ iff $M \vDash \alpha[t]$, for every wff α and time $t \in T_M$.

The importance of this manoeuvre is that it links up the model theory for event logic with that of tense logic; and the latter is a well-researched area with plenty of definitive results for us to draw upon.

3 THE PROOF THEORY OF EVENT LOGIC

The connection established between event logic and tense logic at the end of the previous section is particularly useful because it enables us to specify proof theories for event logic as extensions of already available proof theories for tense logic. In particular, since for an event-logical model $M = \langle T, <, I \rangle$ the temporal frame $\langle T, < \rangle$ is always linear, the tense-logical model $\langle T, <, V_I \rangle$ generated by M will also always be linear. Now it is well-known that the class $\mathcal{L}$ of linear tense-logical models characterises the axiomatic tense-logical system $\mathbf{CL}^7$ defined by the axiom schemata:

(T1) Any tautology of Propositional Calculus
(T2) $G(\alpha \rightarrow \beta) \rightarrow (G\alpha \rightarrow G\beta)$
(T3) $H(\alpha \rightarrow \beta) \rightarrow (H\alpha \rightarrow H\beta)$
(T4) $PG\alpha \rightarrow \alpha$
(T5) $FH\alpha \rightarrow \alpha$
(T6) $G\alpha$ where α is any axiom
(T7) $H\alpha$, where α is any axiom
(T8) $FF\alpha \rightarrow F\alpha$
(T9) $F\alpha \wedge F\beta \rightarrow F(\alpha \wedge \beta) \vee (F(\alpha \wedge F\beta) \vee F(F\alpha \wedge \beta))$
(T10) $P\alpha \wedge P\beta \rightarrow P(\alpha \wedge \beta) \vee (P(\alpha \wedge P\beta) \vee P(P\alpha \wedge \beta))$

and the rule of inference Modus Ponens: from $\alpha \rightarrow \beta$ and α, infer β. For this

---

7. For the system **CL**, see Cocchiarella (1965). More widely available accounts may be found in Prior (1967) and McArthur (1976). The completeness proof for **CL** is outlined in the latter work.

reason, we shall base our Proof Theory for Event Logic on the system **CL** of Linear Tense Logic.

The axioms (T9) and (T10) are rather cumbersome to work with, and for this reason it is often convenient to use instead the following weaker versions of them:

(T9w) $PF\alpha \rightarrow [P\alpha \vee (\alpha \vee F\alpha)]$
(T10w) $FP\alpha \rightarrow [P\alpha \vee (\alpha \vee F\alpha)]$

It is important to realise that (T9w) and (T10w) are strictly weaker than (T9) and (T10): the longer axioms are required for **CL**. But in what follows, the weaker propositions, which are consequences of the stronger ones, will suffice.[8]

All the event logics we shall be considering, then, will include the axioms (T1)–(T10) and the rule Modus Ponens in their proof theories. In addition, they will contain the following axioms, in which 'E' stands for any event radical:

(E1) $ProsE \rightarrow HProsE$
(E2) $PerfE \rightarrow GPerfE$
(E3) $ProsE \rightarrow FPerfE$
(E4) $PerfE \rightarrow PProsE$
(E5) $ProgE \rightarrow PProsE$
(E6) $ProgE \rightarrow FPerfE$
(E7) $ProgE \rightarrow H(ProsE \vee ProgE)$
(E8) $ProgE \rightarrow G(PerfE \vee ProgE)$
(E9) $PProsE \rightarrow [PerfE \vee (ProgE \vee ProsE)]$

The system defined by the axioms (E1)–(E9) is called *Minimal Event Logic*, denoted **E**. Note that the temporal mirror-image of (E9), namely

$$FPerfE \rightarrow [(PerfE \vee ProgE) \vee ProsE]$$

need not be given as an axiom; it is already a theorem of **E**, the proof, somewhat abbreviated, being as follows:

1. $FPerfE \rightarrow FPProsE$	(E4, RF)
2. $FPProsE \rightarrow [PProsE \vee (ProsE \vee FProsE)]$	(T10w)
3. $PProsE \rightarrow [PerfE \vee (ProgE \vee ProsE)]$	(E9)
4. $FProsE \rightarrow FHProsE$	(E1, RF)
5. $FHProsE \rightarrow ProsE$	(T5)
6. $FProsE \rightarrow ProsE$	(4, 5, Tautology)
7. $FPerfE \rightarrow [(PerfE \vee ProgE) \vee ProsE]$	(1, 2, 3, 6, Tautologies)

8. The derivation of (T9w) from the **CL** axioms was given by Lemmon, see Prior (1967), pp. 52–53.

Here RF is the **CL**-metatheorem that from '$\alpha \to \beta$' we can infer '$F\alpha \to F\beta$'. A past-tense version RP of this metatheorem also exists, namely that from '$\alpha \to \beta$' we can infer '$P\alpha \to P\beta$'.

Extensions of **E** can be obtained by the addition of further axioms. Some examples we shall consider are:

Punctual Event Logic (**EP**)

EP is obtained from *E* by the addition of the axiom

(P) $\neg ProgE$.

Once-only Event Logic (**EO**)

EO is obtained from **E** by the addition of the axioms

(O1) $\neg(PerfE \wedge \text{ProsE})$
(O2) $\neg(PerfE \wedge ProgE)$
(O3) $\neg(ProsE \wedge ProgE)$.

Durative Event Logic (**ED**)

ED is obtained from **E** by the addition of the axioms

(D1) $ProsE \to FProgE$
(D2) $PerfE \to PProgE$.

Note that the addition of (P) enables one to simplify the complete axiom-set for **EP**. The axioms (E5)–(E8) are no longer needed, as they are trivial consequences of (P), the antecedent in each case being always false; and (E9) can be simplified by omitting the middle disjunct. It thus becomes possible to define **EP** in a language which has been impoverished to the extent of having the operator *Prog* removed. In such a language the axioms of **EP** can be stated as

(EP1) $ProsE \to HProsE$
(EP2) $PerfE \to GPerfE$
(EP3) $ProsE \to FPerfE$
(EP4) $PerfE \to PProsE$
(EP5) $PProsE \to (PerfE \vee ProsE)$.

This new version of **EP** agrees with the old one in its evaluation of all *Prog*-free wffs; it differs from it only in that it provides no evaluation of wffs containing *Prog*.

A degree of simplification is also possible in the case of **ED**, for in this system (E3) and (E4) can be proved from the other axioms. The proof of (E3) is as follows:

1. $ProsE \rightarrow FProgE$ (D1)
2. $FProgE \rightarrow FFPerfE$ (E6, RF)
3. $FFPerfE \rightarrow FPerfE$ (T8)
4. $ProsE \rightarrow FPerfE$ (1, 2, 3, Tautologies)

(E9) is also redundant in **ED**, since we have the derivation:

1. $PProsE \rightarrow PFProgE$ (D1, RP)
2. $PFProgE \rightarrow [PProgE \vee (ProgE \vee FProgE)]$ (T9w)
3. $PProgE \rightarrow PG(PerfE \vee ProgE)$ (E8, RP)
4. $PG(PerfE \vee ProgE) \rightarrow (PerfE \vee ProgE)$ (T4)
5. $FProgE \rightarrow FH(ProsE \vee ProgE)$ (E7, RF)
6. $FH(ProsE \vee ProgE) \rightarrow (ProsE \vee ProgE)$ (T5)
7. $PProsE \rightarrow [PerfE \vee (ProgE \vee ProsE)]$ (1, 2, 3, 4, 5, 6, Tautologies)

The axioms of **ED** can thus be reformulated as follows:

(ED1) $ProsE \rightarrow HProsE$
(ED2) $PerfE \rightarrow GPerfE$
(ED3) $ProsE \rightarrow FProgE$
(ED4) $PerfE \rightarrow PProgE$
(ED5) $ProgE \rightarrow PProsE$
(ED6) $ProgE \rightarrow FPerfE$
(ED7) $ProgE \rightarrow H(ProsE \vee ProgE)$
(ED8) $ProgE \rightarrow G(PerfE \vee ProgE)$.

For any proof system **S** we shall write $\vdash_S \alpha$ to mean, as usual, that α is provable as a theorem in the system **S**, and $\Sigma \vdash_S \alpha$ to mean that α is deducible from Σ in **S**. The metatheory of Event Logic connects the proof theories and the model theories by showing that the classes of models $\mathcal{M}_E$, $\mathcal{M}_{EP}$, $\mathcal{M}_{EO}$, and $\mathcal{M}_{ED}$ characterize the theories **E**, **EP**, **EO** and **ED** respectively, where as usual to say that a class of models $\mathcal{M}$ *characterizes* a proof system **S** is to say that for every Σ, α in the language under consideration, $\Sigma \vdash_S \alpha$ iff $\Sigma \vDash_{\mathcal{M}} \alpha$. The next two sections are devoted to proving these results.

4 PROOFS OF SOUNDNESS

We treat the two directions of the characterization equivalence separately, namely as:

(1) *Soundness.* If $\Sigma \vdash_S \alpha$ then $\Sigma \vDash_{\mathcal{M}} \alpha$.
(2) *Completeness.* If $\Sigma \vDash_{\mathcal{M}} \alpha$ then $\Sigma \vdash_S \alpha$.

In this section, we shall prove soundness, leaving completeness to Section 5.

We begin with Minimal Event Logic, taking our start from the result already mentioned, that Linear Tense Logic **CL** is sound with respect to the class $\mathscr{L}$ of all linear tense-logical models. This means, *a fortiori*, that **CL** is sound with respect to the class of tense-logical models generated by event-logical models, i.e. the class $\mathscr{L}_E = \{gen(M)|M \in \mathscr{M}_E\}$. This is because all such models are linear, i.e. $\mathscr{L}_E \in \mathscr{L}$. We thus have

$$\vDash_{\mathscr{L}_E} \alpha \text{ iff } M \vDash \alpha[t] \text{ for every } M \in \mathscr{L}_E \text{ and } t \in T_M$$

$$\text{iff } gen(M) \vDash \alpha[t] \text{ for every } M \in \mathscr{M}_E \text{ and } t \in T_M$$

$$\text{iff } M \vDash \alpha[t] \text{ for every } M \in \mathscr{M}_E \text{ and } t \in T_M$$

$$\text{iff } \vDash_{\mathscr{M}_E} \alpha$$

Since **CL** is sound with respect to $\mathscr{L}_E$, the axioms and rules of inference of **CL** are all valid in $\mathscr{L}_E$, and hence, by the above, in $\mathscr{M}_E$. It therefore only remains to show that the event-logical axioms (E1)–(E9) of **E** are also all valid in $\mathscr{M}_E$. We need only consider the odd-numbered axioms, as the proof of validity for each even-numbered axiom is exactly parallel, *mutatis mutandis*, to the proof for the immediately preceding odd-numbered axiom. '*Mutatis mutandis*' here means 'by application of the "Mirror-Image Rule", i.e. by swapping P, H, *Perf*, B, and $x < y$ with F, G, *Pros*, A, and $y < x$, respectively'.

(E1) *ProsE →HProsE*

It suffices to show that $M \vDash HProsE[t]$ for every $M \in \mathscr{M}_E$ such that $M \vDash ProsE[t]$. Suppose $M \vDash ProsE[t]$. This means that $t \in B$ for some $\langle B, A \rangle \in I_M(E)$. Let $t' < t$. Then $t' \in B$ by (FO2). Hence $M \vDash ProsE[t']$. This holds for any $t' < t$, so $M \vDash HProsE[t]$, as required.

(E3) *ProsE → FPerfE*

We must show that if $M \vDash ProsE[t]$, where $M \in \mathscr{M}_E$, then $M \vDash FPerfE[t]$. Suppose $M \vDash ProsE[t]$. Then $t \in B$ for some $\langle B, A \rangle \in I_M(E)$. Since A is non-empty, from the definition of a formal occurrence, there exists some $t' \in A$, so $M \vDash PerfE[t']$. Since $t \in B$ and $t' \in A$, $t < t'$ by (FO1). Hence $M \vDash FPerfE[t]$, as required.

(E5) *ProgE → PProsE*

Suppose $M \vDash ProgE[t]$. Then $t \in c(B) \cap c(A)$ for some $\langle B, A \rangle \in I_M(E)$. Since B is non-empty, let $t' \in B$. Then $M \vDash ProsE[t']$. Also, $t \neq t'$ (since $t \in c(B)$ and $t' \in B$), and $t \nless t'$ (else $t \in B$ by (FO2)). Hence $t' < t$, by linearity, and since $M \vDash ProsE[t']$, we have $M \vDash PProsE[t]$, as required.

(E7) $ProgE \rightarrow H(ProsE \vee ProgE)$

Suppose $M \vDash ProgE[t]$ for some $M \in \mathscr{M}_E$. Then $t \in c(B) \cap (A)$ for some $\langle B, A \rangle \in I_M(E)$. Let $t' \in T_M$ be such that $t' < t$ (such must exist, since any member of B, which is non-empty, must precede t). Then $t' \notin A$, else $t \in A$ by (FO3). Hence either $t' \in c(B) \cap (A)$ or $t' \in B$. In the former case we have $M \vDash ProgE[t']$, in the latter case $M \vDash ProsE[t']$. In either case, therefore, $M \vDash ProsE \vee ProgE[t']$, so since t' is any time preceding t in M, we have $M \vDash H(ProsE \vee ProgE)[t]$, as required.

(E9) $PProsE \rightarrow PerfE \vee (ProgE \vee ProsE)$

Suppose $M \vDash PProsE[t]$. Then for some $t' < t$, $M \vDash ProsE[t']$. This means that $t' \in B$ for some $\langle B, A \rangle \in I_M(E)$. By set theory, either $t \in B$, $t \in c(B) \cap c(A)$, or $t \in A$. But these three cases give $M \vDash ProsE[t]$, $M \vDash ProgE[t]$, and $M \vDash PerfE[t]$ respectively. So in any case we have $M \vDash PerfE \vee (ProgE \vee ProsE)[t]$, as required.

We have now shown that all the axioms and rules of inference of **E** are valid in $\mathscr{M}_E$, and this is sufficient to show that **E** is sound with respect to $\mathscr{M}_E$. It is now a straightforward matter to extend this result to the other systems under consideration, as follows.

Punctual Event Logic

This has the additional axiom (P) $\neg ProgE$. This is not valid in $\mathscr{M}_E$, but it is valid in $\mathscr{M}_{EP}$, since the latter class contains only punctual models. In a punctual model, every occurrence $\langle B, A \rangle$ of every event is such that $B \cup A = T$, i.e. $c(B) \cap c(A) = \varnothing$. Hence we always have that $t \notin c(B) \cap c(A)$, i.e.

$$M \vDash \neg ProgE[t].$$

Once-only Event Logic

This has the further axioms (O1)–(O3), which are again not valid in $\mathscr{M}_E$, but they are valid in $\mathscr{M}_{EO}$, which contains only once-only models. In such a model, $I(E)$ is at most one-membered for every E. Suppose $M \vDash PerfE[t]$, where $M \in \mathscr{M}_{EO}$. Then $t \in A$ for the unique $\langle B, A \rangle \in I(E)$; hence, since $B \cap A = \varnothing$, $t \notin B$ for any $\langle B, A \rangle \in I(E)$, i.e. $M \vDash \neg ProsE[t]$; and also $t \notin c(B) \cap c(A)$ for any $\langle B, A \rangle \in I(E)$, i.e. $M \vDash ProgE[t]$. Hence we have $M \vDash PerfE \rightarrow \neg ProsE[t]$ and $M \vDash PerfE \rightarrow \neg ProgE[t]$, which are straightforwardly equivalent to $M \vDash \neg (PerfE \wedge ProsE)[t]$ and

$M \vDash \neg(PerfE \wedge ProgE)[t]$ respectively, thereby validating (O1) and (O2). (O3) is validated similarly.

Durative Event Logic

This has the extra axioms (D1) and (D2). We need only show validity for (D1), the proof for (D2) being *mutatis mutandis* exactly similar. Suppose, then, that $M \vDash ProsE[t]$ for some $M \in \mathcal{M}_{ED}$ and $t \in T_M$. Then $t \in B$ for some $\langle B, A \rangle \in I_M(E)$. Since M is a durative model, $c(B) \cap c(A) \neq \varnothing$, so let $t' \in c(B) \cap c(A)$. Then $M \vDash ProgE[t']$. Since $t \in B$ and $t' \in c(B)$, $t \neq t'$; also $t' \not< t$ (else $t' \in B$ by (FO2)); hence $t < t'$. Since $M \vDash ProgE[t']$, we have $M \vDash FProgE[t]$, as required. This completes the soundness proofs.

5 PROOFS OF COMPLETENESS

We begin by considering quite generally what is required for a proof that a given proof theory **S** is complete with respect to a given class of models $\mathcal{M}$. The desired result, namely

$$\text{'If } \Sigma \vDash_{\mathcal{M}} \alpha \text{ then } \Sigma \vdash_S \alpha\text{'},$$

is first rewritten as

$$\text{'If } \Sigma \nvdash_S \alpha \text{ then } \Sigma \nvDash_{\mathcal{M}} \alpha\text{'}$$

which in turn is rewritten as

$$\text{'If } \Sigma \cup \{\neg\alpha\} \text{ is consistent in S then } \Sigma \cup \{\neg\alpha\} \text{ has a model in } \mathcal{M}\text{'}$$

or, more tidily, as

$$\text{'Any set of wffs consistent in S has a model in } \mathcal{M}\text{'}.$$

By a model *for* a set of wffs Σ, in the case of tense- and event-logical models, is meant a model M such that for some $t_0 \in T_M$, $M \vDash \sigma[t_0]$ for each $\sigma \in \Sigma$.

For Minimal Event Logic, then, we must show there exists a model in $\mathcal{M}_E$ for any set of wffs consistent in **E**. Let Σ be our consistent set, and extend Σ in the usual way to a maximally consistent set Σ^+, i.e. a consistent extension of Σ to which no further wffs may be added without incurring inconsistency. Now note that if Σ is consistent in **E** then it is consistent in **CL**, since every derivation in **CL** is also a derivation in **E**, so any proof of inconsistency in **CL** would also be a proof of inconsistency in **E**. Since **CL** is complete with respect to $\mathcal{L}$, any set Σ consistent in **E** has a model M in $\mathcal{L}$ (for the purposes of this model, wffs of the forms $PerfE$, $ProgE$, and $ProsE$ are considered as unanalysed primitives). We shall show how to derive from this model a model $M' \in \mathcal{M}_E$ such that $M = gen(M')$. We do this by showing how to specify the value of $I(E)$ for each radical E occurring in Σ^+.

Assume, then, that $M \vDash \alpha[t_0]$ for each $\alpha \in \Sigma^+$. Since Σ^+ is maximally consistent, we can actually assert the stronger result that $M \vDash \alpha[t_0]$ iff $\alpha \in \Sigma^+$. Then, given a radical E, define the sets

$$PERF = \{t \in T_M | M \vDash Perf\,E[t]\}$$

$$PROS = \{t \in T_M | M \vDash Pros\,E[t]\}$$

$$PROG = \{t \in T_M | M \vDash Prog\,E[t]\}.$$

Now let α be any theorem of E. Since Σ^+ is maximally consistent in E, $\alpha \in \Sigma^+$, and hence $M \vDash \alpha[t_0]$. By axioms (T6) and (T7), $G\alpha$ and $H\alpha$ are also theorems of E, so $M \vDash G\alpha[t_0]$ and $M \vDash H\alpha[t_0]$. Since M is a CL-model, these imply that $M \vDash \alpha[t]$ for all $t > t_0$ and all $t < t_0$ respectively. Hence $M \vDash \alpha[t]$ for all $t \in T_M$.

In particular, suppose $t \in PROS$. Then $M \vDash Pros\,E[t]$, and since (E1) is an axiom of E, $M \vDash ProsE \to HProsE[t]$. Hence $M \vDash HProsE[t]$, i.e. $M \vDash Pros\,E[t]$ for all $t' < t$, i.e. $t' \in PROS$ for all $t' < t$. Hence $PROS$ is an initial segment of T, i.e. any time which precedes a member of $PROS$ is itself a member of $PROS$. Similarly, it can be shown that $PERF$ is a final segment of T, the proof in this case making use of axiom (E2).

Again, if $t \in PROS$, so $M \vDash Pros\,E[t]$, then from axiom (E3) $M \vDash FPerf\,E[t]$, so for some $t' \in T$, $t < t'$ and $M \vDash Perf\,E[t']$, so $t' \in PERF$. Hence if $PROS$ is non-empty, so is $PERF$. Similarly, (E4) can be used to show that if $PERF$ is non-empty then so is $PROS$. Combining these results gives us that $PERF = \varnothing$ iff $PROS = \varnothing$.

If $t \in PROG$, then $M \vDash Prog\,E[t]$, so from axioms (E5) and (E6), $M \vDash PPros\,E[t]$ and $M \vDash FPerf\,E[t]$. The first of these guarantees the existence of a $t' \in T$ with $M \vDash Pros\,E[t']$, so $t' \in PROS$; and the second likewise gives us a $t'' \in PERF$. So if $PROG \neq \varnothing$, then $PROS \neq \varnothing$ and $PERF \neq \varnothing$.

Now assume $PROG \neq \varnothing$ (so $PROS \neq \varnothing$), and let $t \in T$. Then so long as t is not the first member of T, it is preceded by some $t' \in PROS$, so $M \vDash PPros\,E[t]$. By axiom (E9), this means that $M \vDash PerfE \vee (ProgE \vee ProsE)[t]$ and hence that $t \in PROS \cup PROG \cup PERF$. If t is not the last member of T, the same result follows using the mirror image of (E9), which we proved as a theorem. Finally, if t is the *only* member of T, then (E3), (E4) and (E5) respectively imply that $M \vDash \neg Pros\,E[t]$, $M \vDash \neg Perf\,E[t]$, and $M \vDash \neg Prog\,E[t]$, so in this case $PROS = PERF = PROG = \varnothing$. In conclusion, then, we have that if $PROS$ and $PERF$ are non-empty, then $PROS \cup PROG \cup PERF = T$.

Now for a given radical E, let us say that a formal occurrence $\langle B, A \rangle$ in $\langle T, < \rangle$ is *compatible*[9] with E so long as $A \subseteq PERF$, $B \subseteq PROS$, and

---

9. I am indebted to Kit Fine for suggesting the idea of compatible occurrences, thereby enabling me to simplify the proof of completeness considerably.

$c(B) \cap c(A) \subseteq PROG$. We shall show that

$$PERF = \bigcup \{A \mid \langle B, A \rangle \text{ is compatible with } E\}$$

$$PROS = \bigcup \{B \mid \langle B, A \rangle \text{ is compatible with } E\}$$

$$PROG = \bigcup \{c(B) \cap c(A) \mid \langle B, A \rangle \text{ is compatible with } E\}$$

In fact, it follows immediately from the definition of compatibility that the right-hand side of each of these equalities is included in the left-hand side. It remains, therefore, to show that the reverse inclusions also hold.

First, let $t \in PERF$. We must show that $t \in A$ for some $\langle B, A \rangle$ compatible with E. We distinguish two cases:

(a) Suppose $PROS \cup PERF = T$. In this case, consider the formal occurrence $\langle B, A \rangle$, where $B = c(PERF)$ and $A = PERF$. Since $B = c(PERF) \subseteq PROS$, $A \subseteq PERF$, and $c(B) \cap c(A) = PERF \cap c(PERF) = \varnothing \subseteq PROG$, $\langle B, A \rangle$ is compatible with E, and moreover $t \in PERF = A$, as required.

(b) Suppose $PROS \cup PERF \neq T$. In this case, consider the formal occurrence $\langle B, A \rangle$, where $B = PROS$ and $A = PERF$. Since $B \subseteq PROS$, $A \subseteq PERF$, and $c(B) \cap c(A) = c(PROS) \cap c(PERF) \subseteq PROG$ (since $PROS \cup PROG \cup PERF = T$), $\langle B, A \rangle$ is compatible with E, and $t \in PERF = A$ as required.

This proves the first equality; and the second can be proved exactly similarly, *mutatis mutandis*.

For the third equality, let $t \in PROG$. We must show that $t \in c(B) \cap c(A)$ for some $\langle B, A \rangle$ compatible with E. Define $\text{Bf}(t) = \{t' \in T \mid t' < t\}$ (the set of times *before* t), and $\text{Af}(t) = \{t' \in T \mid t < t'\}$ (the set of times *after* t), and consider the formal occurrence $\langle B, A \rangle$, where $B = PROS \cap \text{Bf}(t)$ and $A = PERF \cap \text{Af}(t)$. Then $B \subseteq PROS$, $A \subseteq PERF$, and

$$c(B) \cap c(A) = c[PROS \cap \text{Bf}(t)] \cap c[PERF \cap \text{Af}(t)]$$

$$= [c(PROS) \cup c(\text{Bf}(t))] \cap [c(PERF) \cup c(\text{Af}(t))]$$

$$= [c(PROS) \cup \text{Af}(t) \cup \{t\}] \cap [c(PERF) \cup \text{Bf}(t) \cup \{t\}]$$

$$= [c(PROS) \cap c(PERF)] \cup [c(PROS) \cap \text{Bf}(t)] \cup [c(PERF) \cap \text{Af}(t)] \cup \{t\}.$$

Now $c(PROS) \cap c(PERF) \subseteq PROG$ (as in (b) above), and $\{t\} \subseteq PROG$. Also, by axiom (E7), since $M \vDash ProgE[t]$, we have $M \vDash ProsE \lor ProgE[t']$ for all $t' \in \text{Bf}(t)$, i.e. $\text{Bf}(t) \subseteq PROS \cup PROG$. Hence $c(PROS) \cap \text{Bf}(t) \subseteq PROG$. Similarly, using axiom (E8), $c(PERF) \cap \text{Af}(t) \subseteq PROG$. But a union of subsets of $PROG$ is again a subset of $PROG$, so $c(B) \cap c(A) \subseteq PROG$. Hence $\langle B, A \rangle$ is compatible with E, and also $t \in c(B) \cap c(A)$, as required.

Let us now for each radical E define

$$I(E) = \{\langle B, A \rangle | \langle B, A \rangle \text{ is compatible with } E\}.$$

Define an event-logical model $M' = \langle T, <, I \rangle$. We then have that

$M' \vDash PerfE[t]$ iff $t \in A$ for some $\langle B, A \rangle \in I(E)$

                   iff $t \in A$ for some $\langle B, A \rangle$ compatible with E

                   iff $t \in PERF$

                   iff $M \vDash PerfE[t]$

and similarly

$$M' \vDash ProsE[t] \text{ iff } M \vDash ProsE[t]$$

and

$$M' \vDash ProgE[t] \text{ iff } M \vDash ProsE[t].$$

Since M and M' agree on all the atomic wffs, and have the same recursive component in their truth-definitions, it follows that they have the same theory (in fact $M = gen(M')$). Hence, since $M \vDash \alpha[t_0]$ for each $\alpha \in \Sigma^+$, $M' \vDash \alpha[t_0]$ for each $\alpha \in \Sigma^+$ also, so M' is a model for Σ, as required.

We have now proved that **E** is complete with respect to $\mathcal{M}_E$. The extension of this result to the other event logics we have considered is straightforward, as follows.

Punctual Event Logic

We have, for every E and t, $M \vDash \neg ProgE[t]$. Hence $PROG = \varnothing$. This means that the formal occurrences $\langle B, A \rangle$ which are compatible with E all have $c(B) \cap c(A) = \varnothing$, i.e. they are all punctual. Hence the model M' constructed in the previous proof is automatically punctual in this case, i.e. belongs to $\mathcal{M}_{EP}$. Hence **EP** is complete with respect to $\mathcal{M}_{EP}$.

Once-only Event Logic

For every E and t we have $M \vDash \neg(PerfE \wedge ProsE)[t]$, $M \vDash \neg(PerfE \wedge ProgE)[t]$, and $M \vDash \neg(ProsE \wedge ProgE)[t]$. These imply that $PROS \cap PERF = PERF \cap PROG = PROS \cap PROG = \varnothing$; hence the only formal occurrence compatible with E is $\langle PROS, PERF \rangle$. So $I(E)$ must be $\{\langle PROS, PERF \rangle\}$, and M' is therefore once-only, i.e. belongs to $\mathcal{M}_{EO}$. Hence **EO** is complete with respect to $\mathcal{M}_{EO}$.

Durative Event Logic

We cannot show that the occurrences compatible with E are all durative; they need not be. Instead, we show that if we put

$$I(E) = \{\langle B, A\rangle \mid \langle B, A\rangle \text{ is durative and compatible with } E\}$$

then everything goes through as required. What is needed is to show that, in **ED**, we have for each E

$$PERF = \bigcup\{A \mid \langle B, A\rangle \text{ is durative and compatible with } E\}$$

$$PROS = \bigcup\{B \mid \langle B, A\rangle \text{ is durative and compatible with } E\}$$

$$PROG = \bigcup\{c(B) \cap c(A) \mid \langle B, A\rangle \text{ is durative and compatible with } E\}.$$

As before, the inclusion of the right-hand side in the left-hand side is immediate, so we need only consider the reverse inclusion.

Let $t \in PERF$, so $M \vDash PerfE[t]$. By axiom (D2), $M \vDash PProgE[t]$, so there is a time $t' < t$ such that $M \vDash ProgE[t']$. By axiom (E7), for any time t'' preceding t', either $M \vDash ProgE[t'']$ or $M \vDash ProsE[t'']$. Hence $Bf(t') \subseteq PROS \cup PROG$, so $c(PROS) \cap Bf(t') \subseteq PROG$. Let $B = PROS \cap Bf(t')$, so $B \subseteq PROS$. Similarly, by axiom (E8), $Af(t') \subseteq PERF \cup PROG$, so $c(PERF) \cap Af(t') \subseteq PROG$. Let $A = PERF \cap Af(t')$, so $A \subseteq PERF$ and $t \in A$. Then as above, $c(B) \cap c(A) \subseteq PROG$, and $\langle B, A\rangle$ is compatible with E. Since $t' \in (B) \cap c(A)$, $\langle B, A\rangle$ is durative. But $t \in A$, so

$$t \in \bigcup\{A \mid \langle B, A\rangle \text{ is durative and compatible with } E\}$$

as required.

The second inclusion can be proved in an exactly similar way, *mutatis mutandis*. The third inclusion follows immediately from the corresponding inclusion in the general case, since we have

$$\bigcup\{c(B) \cap c(A) \mid \langle B, A\rangle \text{ is compatible with } E\}$$

$$= \bigcup\{c(B) \cap c(A) \mid \langle B, A\rangle \text{ is durative and compatible with } E\}.$$

This is because in the left-hand side of this equation, the only pairs $\langle B, A\rangle$ which contribute to the union are the durative ones, since by definition a $\langle B, A\rangle$ for which $c(B) \cap c(A) \neq \varnothing$ is durative.

We have now proved that **ED** is complete with respect to $\mathcal{M}_{ED}$, and all the promised soundness and completeness proofs have been duly delivered.

6   REPRESENTATION OF TEMPORAL REFERENCE

So far, I have presented a series of extensions of tense logic designed to facilitate reasoning about the occurrence of events. All these extensions are shown to be sound and complete with respect to appropriate model theories. The systems presented here have the following limitations:

(a) No mechanism is incorporated for explicitly referring to individual times;
(b) No account is given of the internal structure of events.

In this section and the next I shall indicate how these limitations might be overcome within the framework of Event Logic.

The first problem to be surmounted is how to express that an event occurs at a particular time, not just sometime in the past or sometime in the future. In order to do this it is necessary to find a way of incorporating explicit temporal reference into the system. The approach which I favour is to take the basic form of temporal reference to be propositions such as 'It is five o'clock', 'It is January', 'It is 1985', and so on. Such propositions may be called *chronological propositions*. Ultimately, we can expect them to be analysed in terms of propositions about clocks, calendars, astronomical phenomena, etc. We may provisionally divide chronological propositions into *instant propositions* and *period propositions*, the former (like 'It is five o'clock') being true at isolated instants, the latter (like 'It is January') over periods; but later on, this classification will have to be reconsidered.

For assigning dates to states, the apparatus of tense logic now suffices, so that propositions of the form 'α at i' (where i is an instant) are analysed as $P(i \wedge \alpha)$, $i \wedge \alpha$, or $F(i \wedge \alpha)$, according as i is past, present, or future; while propositions of the forms 'α throughout p' and 'α during p' (where p is a period) receive analyses like $H(p \to \alpha)$ and $P(p \wedge \alpha)$ respectively (for p in the past), or $G(p \to \alpha)$ and $F(p \wedge \alpha)$ respectively (for p in the future).

In order to assign dates to events, the language must be augmented with a new operator WHILE such that, for a radical E and a proposition α, WHILE(E, α) is a radical whose occurrences are just those occurrences of E which occur at a time when α is true. The semantic rule for WHILE is

$$I_M(\text{WHILE}(E, \alpha)) =$$

$$\{\langle B, A \rangle \in I_M(E) | \exists b \in B : \exists a \in A : \forall t : b \leqslant t \leqslant a \Rightarrow M \vDash \alpha[t]\}.$$

Note that the meaning of WHILE thus given only covers a portion of the meaning of the English temporal 'while'; it is the 'while' of

While he was in Paris, John visited the Louvre

(i.e. *Perf*(WHILE(*John-VISIT-the Louvre, John is in Paris*))) rather than the 'while' joining two states in

While John was in Paris, Mary was in Rome

(i.e. $H(p \wedge John\ is\ in\ Paris \rightarrow Mary\ in\ in\ Rome)$, where p is a proposition which determines which occasion of John's visiting Paris is being referred to), or the 'while' in which a state is terminated by an event, as in

Mozart died while composing his Requiem.

Despite this restricted interpretation, I believe that the WHILE defined as above should suffice for most purposes.

We can now analyse propositions of the form 'E happened during p', where p is a period, as $Perf(\mathrm{WHILE}(E, p))$. The case where a punctual event is located at an instant is trickier, and obviously cannot be analysed in terms of WHILE. This case in fact forces us to reconsider the nature of our instant propositions. Because of the way punctual events are defined in Event Logic, when a punctual event happens there is no element of T at which it happens; rather it happens in the 'gap' between a B-set and an A-set, where $B \cup A = T$. If we wish to hold on to the idea that the location of an instantaneous event is an instant, then we cannot regard the members of T as instants; at best, they will be short periods, and if we consider them to be indivisible (as we must, in view of the linear ordering imposed on T), then this must be because *nothing happens* during them, i.e. all change that is visible in the model occurs in the 'gaps' between time-atoms. It is these gaps (characterized in our model as pairs of the form $\langle B, A \rangle$, where $B \cup A = T$) that play the part of instants in this concept of time.

What, now, of instant propositions? Clearly, they can no longer *be* propositions, since propositions are true or false on elements of T, not in the gaps between them. Instead, we introduce instant chronological *events*: so that 'five o'clock' is now something that *happens*, not something that is the case. We must represent it not by a proposition 'It is five o'clock', but by a *radical* 'It-BE-five-o'clock'. In fact, an instant i is represented by the punctual event E such that $I(E) = \{\langle B, A \rangle\}$, where B is the set of times preceding i and A is the set of times following i (note that i itself is not a member of T). This is a once-only event, of course, but if we are thinking of five o'clock, say, in the generic sense, as something that happens twice every day, then we need only consider the repeatable event whose occurrences are all the individual five o'clock occurrences.

A sentence like 'It was raining at five o'clock' must now be analysed not as

P(It is raining $\wedge$ It is five o'clock)

but as

Perf(WHILE(It-BE-five-o'clock,   It is raining))

which is analogous to 'The clock struck five while it was raining' (= 'It was raining when the clock struck five').

In order now to say that an instantaneous event occurred at an instant, e.g. 'It started raining at five o'clock', we need to find a way of expressing the simultaneous occurrence of two punctual events. It appears not to be possible to do this within the existing formalism; a new operator SIMUL must be introduced such that for punctual events E_1 and E_2, $SIMUL(E_1, E_2)$ is again a punctual event, the semantic rule being straightforward:

$$I(SIMUL(E_1, E_2)) = I(E_1) \cap I(E_2).$$

We can now express propositions of the form 'E happened at i', where E is a punctual radical and i is an instant, as

$$Perf(SIMUL(E, i)).$$

I have given here only a brief outline of how temporal reference can be incorporated into Event Logic. It is not an easy matter to handle, and I have not yet worked out a fully-fledged theory, but it seems likely that some solution along the lines sketched here should be possible. Philosophically speaking, it is clear that ultimately it must be possible to analyse explicit temporal reference in terms of the most basic temporal concepts, those concerning simultaneity and succession.

This is perhaps also the place to mention a further possibility for syntactic innovation in event logic. Just as tense logic was enriched by the addition of Kamp's 'since' and 'until' operators S and U (Kamp, 1968), so too we may consider the introduction of similar operators into event logic, enabling us to formalize sentences like

$$\textit{John will be in the garden until} \begin{cases} \textit{Mary arrives} \\ \textit{five o'clock.} \end{cases}$$

$$\textit{John has been in the garden since} \begin{cases} \textit{Mary arrived.} \\ \textit{five o'clock.} \end{cases}$$

in which a state is brought into temporal relationship with an event rather than another state as would be required for Kamp's operators to be applied.

We may posit, then, *binary aspect operators* SINCE and UNTIL such that, for a proposition α and a radical E, $SINCE(\alpha, E)$ and $UNTIL(\alpha, E)$ are both propositions, with meanings defined by the rules:

$M \models SINCE(\alpha, E)[t]$ iff there is $\langle B, A \rangle \in I_M(E)$ such that $t \in A$ and

$$M \models \alpha[t'] \text{ for every } t' \in A \cap Bf(t).$$

$M \models UNTIL(\alpha, E)[t]$ iff there is $\langle B, A \rangle \in I_M(E)$ such that $t \in B$ and

$$M \models \alpha[t'] \text{ for every } t' \in B \cap Af(t).$$

From these semantic rules, we can easily derive the consequences

$$\vDash SINCE(\alpha, E) \rightarrow PerfE$$

$$\vDash UNTIL(\alpha, E) \rightarrow ProsE.$$

As I have remarked elsewhere in connection with Kamp's operators (this volume, p. 12), the second of these implications makes UNTIL unsuitable as a representation of many uses of 'until' in English. For this reason, it would be helpful also to introduce *weak* versions of SINCE and UNTIL (cf. p. 40) defined by the rules:

$$SINCE^*(\alpha, E) =_{def} SINCE(\alpha, E) \lor H\alpha$$

$$UNTIL^*(\alpha, E) =_{def} UNTIL(\alpha, E) \lor G\alpha.$$

As a first approximation, we may say that the English connectives 'since' and 'until' correspond to our SINCE and UNTIL* respectively.

Armed with the strong and weak binary aspect operators, we can define other tense and aspect operators as follows:

$$PerfE =_{def} SINCE(true, E)$$

$$ProsE =_{def} UNTIL(true, E)$$

$$H\alpha =_{def} SINCE^*(\alpha, null)$$

$$G\alpha =_{def} UNTIL^*(\alpha, null).^{10}$$

There does not appear, however, to be any way of defining *Prog* in terms of these operators. Nor does it appear possible to define Kamp's S and U in terms of them, although we have the following implications:

$$\alpha U\beta \rightarrow \beta \lor UNTIL(\alpha, Ingr\beta)$$

$$\alpha S\beta \rightarrow \beta \lor SINCE(\alpha, Ingr\neg\beta).^{11}$$

Turning now to WHILE, we find a clear connection with UNTIL, although it is hard to state it rigorously. For a once-only punctual event E, given that α is true now, to say that α will be true until E happens is to say that

10. Here '*null*' denotes the empty event, i.e. the event with no occurrences, defined semantically by the rule

$$I_M(null) = \varnothing$$

or axiomatically by the rule

$$\vdash \neg Perf\, null.$$

11. For the definition of *Ingr*, see below (p. 194).

α will not become false while E has not yet happened, i.e.

$$\alpha \rightarrow [\text{UNTIL}(\alpha, E) \equiv \neg Pros\text{WHILE}(Ingr\neg\alpha, \neg Perf E)],$$

so that, for example, 'I will wait until you get back' can be paraphrased as 'I won't go while you haven't got back'.[12] This connection between the meanings of 'while' and 'until' is reflected in the uses to which these words are put in computer languages, as illustrated by the near-equivalence (modulo end-effects) of the *Pascal* commands

repeat $\langle procedure \rangle$ **until** $\langle condition \rangle$

and

while not $\langle condition \rangle$ **do** $\langle procedure \rangle$.

Clearly there is much scope for further investigation here.

7   THE INTERNAL STRUCTURE OF EVENTS

By the 'internal structure of events', I mean in particular the decomposition of an event into its constituent states. In saying this, I do not wish to imply that every event can be straightforwardly broken down into a sequence of states of the form 'first α_1, then $\alpha_2, \ldots$, then α_n': the truth is clearly more complicated. What I have in mind is rather that since many events *can* be analysed in terms of states, such analyses ought to be reflected in our formal apparatus.

The two simplest constructs would appear to be the definition of a punctual event as either the *inception* or the *cessation* of some state, and the definition of certain durative events ('atelic' ones) in terms of some state's obtaining *for a while*. In order to model these constructs in Event Logic, we introduce two operators *Ingr* and *Po*, which form radicals out of propositions.

The radical *Ingr*α denotes that event each of whose occurrences consists of the proposition α's becoming true. Semantically, $I(Ingr\alpha)$ is the set of pairs $\langle B, c(B) \rangle$ such that α is false throughout some final subinterval of B and true throughout some initial subinterval of $c(B)$.[13] As logical consequences of this definition we obtain the equivalences

$$\vDash PerfIngr\alpha \equiv P(P\neg\alpha \wedge \alpha) \vee (P\neg\alpha \wedge \alpha)$$

$$\vDash ProsIngr\alpha \equiv (\neg\alpha \wedge F\alpha) \vee F(\neg\alpha \wedge F\alpha)$$

---

12. In Yorkshire, this might come out as 'I will wait while you get back'. And the Leeds man who serviced my car advised me not to wait 'while October' before getting rid of it.
13. 'Subinterval' must here be understood to include singleton subsets of T.

which could accordingly be used to give contextual syntactic definitions of *Ingr* (cf. Galton, 1984, p. 50, where *Ingr* is introduced in this way).

The radical *Po*α denotes that event each of whose occurrences consists of a stretch of α's being uninterruptedly true, preceded and followed by times at which α is false. Semantically, $I(Po\alpha)$ is the set of formal occurrences $\langle B, A \rangle$ such that p is false thoughout some final subinterval of B and throughout some initial subinterval of A, and true throughout $c(B) \cap c(A)$, this last-mentioned set being non-empty. Contextual definitions similar to those for *Ingr* can also be given for *Po* (see Galton, 1984, p. 82).

A further construct which might be introduced is *succession* or *sequential composition* (cf. Allen, 1984, p. 134), the following of one event by another. There are two different versions of succession: *immediate succession*, in which the second event follows 'hot on the heels' of the first, i.e. with no intervening gap; and *delayed succession*, in which a gap is possible. We shall write THEN(E_1, E_2) for this second case, THEN*(E_1, E_2) for the former more restricted case.

The semantic definitions are easy enough to give:

$$I(\text{THEN}(E_1, E_2)) = \{\langle B, A \rangle | \exists B' : \exists A' : B \cup A' = T \wedge$$

$$\langle B, A' \rangle \in I(E_1) \wedge \langle B', A \rangle \in I(E_2)\}$$

$$I(\text{THEN*}(E_1, E_2)) = \{\langle B, A \rangle | \exists B' : \exists A' : B' = c(A) \wedge$$

$$\langle B, A' \rangle \in I(E) \wedge \langle B', A \rangle \in I(E_2)\}.$$

Note that if E_1 and E_2 are punctual, THEN*(E_1, E_2) is the same as SIMUL(E_1, E_2), since the only way that two punctual events can occur with no intervening gap is by being simultaneous. Also note that any occurrence of *Po*α is also an occurrence of THEN(*Ingr*α, *Ingr*($\neg \alpha$)); but the converse does not hold, although so long as α is such that it cannot change truth-value infinitely often in a bounded interval, any occurrence of the latter event must be an occurrence either of *Po*α or of THEN(*Po*α, *Po*α).

As in the previous section, it is necessary to remark that the definitions given here can only be provisional until a fully fledged theory has been worked out. I offer them merely as an indication of the potential of Event Logic as a basis for further elaboration of the logic of events.

References

Allen, J. F. (1984), 'Towards a general theory of action and time', *Artificial Intelligence*, **23**, 123–154.

Cocchiarella, N. B. (1965), 'Tense and Modal Logic: A Study in the Topology of Temporal Reference', PhD Thesis, University of California in Los Angeles.

Davidson, D. (1967), 'The logical form of action sentences', in N. Rescher (ed.), *The Logic of Decision and Action* University of Pittsburgh Press, 81–95, reprinted in D. Davidson (1980), *Essays on Actions and Events* Oxford: Clarendon Press, 105–121.

Dowty, D. (1979). *Word Meaning and Montague Grammar* Dordrecht: D. Reidel.

Galton, A. P. (1984), *The Logic of Aspect* Oxford: Clarendon Press.

Hughes, G. E. and Cresswell, M. J. (1984), *A Companion to Modal Logic* London and New York: Methuen.

Kamp, J. A. W. (1968), 'Tense logic and the theory of linear order', PhD thesis, University of California in Los Angeles.

Kowalski, R. A. and Sergot, M. J. (1986), 'A logic-based calculus of events', *New Generation Computing*, **4**, 67–95.

McArthur, R. P. (1976), *Tense Logic* Synthese Library, Vol. iii. Dordrecht: D. Reidel.

McDermott, D. (1982), 'A temporal logic for reasoning about processes and plans', *Cognitive Science*, **6**, 101–155.

Prior, A. N. (1967), *Past, Present, and Future* Oxford: Clarendon Press.

Tichý, P. (1985), 'Do we need interval semantics?', *Linguistics and Philosophy*, **8**, 263–282.

6 MODAL AND TEMPORAL LOGIC PROGRAMMING

Dov Gabbay

*Imperial College of Science
and Technology,
University of London*

ABSTRACT

We extend Horn clause programming to allow for modal and temporal connectives such as possibility $\Diamond$. If **P** is a program and B is the body of a clause then $B \rightarrow \Diamond \wedge \mathbf{P}$ is also a clause. The computation procedures are given and soundness is proved. It is also shown that computing in modal and temporal logic is like computing in classical logic without skolemizing.

BACKGROUND

We want to study what capabilities Logic Programming can have for handling time and modality. Temporal and modal phenomena seem to arise in an instrinsic way in many branches of Computer Science including those branches in which Logic Programming seems to be most successful. It is therefore necessary to analyse logically the relationship of Logic Programming with Temporal and Modal logics.

There are several ways of presenting a Temporal or a Modal logical system. Main among them are:

(1) presentation in predicate logic and especially in Horn clause logic itself;
(2) semantic presentation via models and special interpretations, for specially chosen temporal connectives;
(3) proof theoretic and axiomatic presentations (Gentzen systems, Tableaux, Hilbert axiom systems etc).

We begin by explaining how these three methods relate to one another, in

Temporal Logics and their Applications

ISBN: 0-12-274060-2

order to set the scene for our proposed system of Temporal and Modal Logic Programming.

Consider a model **M** of the classical predicate calculus involving the predicates $A(x)$, $R(x, y)$ and a domain D. $A(x)$ could mean, for example, 'x is in attendance' and $R(x, y)$ could mean 'x is sitting next to y'. The domain is the set of registered students of my logic class **logic 140**. The particular extensions of $A(x)$ and $R(x, y)$ are chosen by looking at the first day of classes:

$A(x)$ is true iff x attended the first logic lecture;
$R(x, y)$ is true iff x was sitting next to y in the first lecture.

We thus get a model **M1** for the language with $A(x)$ and $R(x, y)$.

Assume now that we record attendance and sitting positions throughout all my lectures. We will get a sequence of models **M1, M2, M3,...**, with different extensions for the predicate A and different extensions for the predicate $R(x, y)$.

This can be described as follows:

3	**M3**	$A(\text{John}) = T$
2	**M2**	$A(\text{John}) = T$
1	**M1**	$A(\text{John}) = F$

On the right hand side we have recorded, for our convenience, the truth value of one particular predicate $A(\text{John})$ (i.e. 'John attends class'). We could have done this for all possible predicates of the language.

Suppose now we want to praise John for his attendance and say that John attended all lectures except the first one. How do we express that? Notice that all we have is a sequence of models; we do not have a suitable language for *talking about* the sequence of models. Here comes the need for a temporal language of one sort or another. We may want to say some complex things about John's attendance, such as that he attended every other class, or to correlate his attendance with the person or persons sitting next to him.

The first two approaches for talking about time can be illustrated in this example. First note that to be able to talk about time we need the earlier-later relation $<$ ('smaller than' in our case, since time is represented by numbers). Talking about time would also involve variables t, s ranging over time; x, y range over D.

How do we join the language with $A(x)$, $R(x, y)$ and the language with t, s, $<$, into one language?

The first approach is to incorporate time **T** and $<$ (in our case

$(\{1, 2, 3, \ldots\}, <))$ into predicate logic itself. Thus we no longer view our situation as a sequence of models **M**1, **M**2, **M**3, ... but push the time $\{1, 2, 3, \ldots\}$ into the predicates and talk about $A^*(1, \text{John})$, $A^*(2, \text{John})\ldots$ $R^*(1, \text{John, Mary})$, $R^*(2, \text{John, Mary})$, ... Thus '$R(x, y)$ is true at model **M**k' becomes '$R^*(k, x, y)$'. Similarly '$A(x)$ is true at model **M**k' becomes '$A^*(k, x)$'. Generally, with any formula $\psi(x, y)$ of the predicate language we can associate a formula $\psi^*(t, x, y)$ saying that $\psi(x, y)$ is true in the model **M**t. ψ^* may contain the earlier-later predicate $<$.

The above *looks like* first-order predicate logic with a variable t ranging over time and the relation $<$. It is not, however, exactly first-order predicate logic. Firstly we do not allow ourselves to write $A^*(8, 10)$ '(the number 10 attended the eighth class)'; nor can we write $A^*(\text{John}, 10)$ or $A^*(\text{John, Mary})$ or John $<$ Mary etc. Furthermore when we quantify $\forall t\, A^*(t, x)$ (read: 'x attended all classes') the quantifier $\forall$ ranges over time **T** only. In $\forall x\, A^*(1, x)$ (read: 'everyone attended the first class') the quantifier $\forall$ ranges over the domain D of students only. Thus we have here a restricted type of predicate logic, with sorts and restricted quantification. We shall see later why this is not exactly first order predicate logic.

The second approach to talking about time is to enable each model **M**k to talk about other models, e.g. **M**$(k + 1)$. To do this we need special additional connectives in the language of predicate logic. For example, if we add to the language of predicate logic besides $\sim$, $\wedge$, $\vee$, $\rightarrow$, $\forall$, and $\exists$ the connective Y, and read YA as 'A was true yesterday' then we can say for example:

3

2 $\quad YA$ (John) $=$ true

1 $\quad A(\text{John}) =$ true

$YA(\text{John})$ is true in the model **M**2 (with the enriched language with Y) iff $A(\text{John})$ is true at **M**1.

The usual temporal connectives one adds to the language of predicate logic are:

$\quad FA =$ it will be the case that A

$\quad PA =$ it has been the case that A

$\quad S(A, B) =$ since A was true B has always been true

$\quad U(A, B) =$ until A is true B will always be true

$\quad GA =$ it will always be the case that A

$\quad HA =$ it has always been the case that A.

The truth conditions for the new connectives are:

FA is true at $\mathbf{M}k$ iff for some $t > k$, A is true at $\mathbf{M}t$.

PA is true at $\mathbf{M}k$ iff for some $t < k$, A is true at $\mathbf{M}t$.

$S(A, B)$ is true at $\mathbf{M}k$ iff for some $t < k$, A is true at $\mathbf{M}t$ and for all s,
$\qquad\qquad\qquad$ if $t < s < k$ then B is true at $\mathbf{M}s$.

$U(A, B)$ is true at $\mathbf{M}k$ iff for some $t > k$, A is true at $\mathbf{M}t$ and for all s,
$\qquad\qquad\qquad$ if $k < s < t$ then B is true at $\mathbf{M}s$.

GA is true at $\mathbf{M}k$ iff for all $t > k$, A is true at $\mathbf{M}t$.

HA is true at $\mathbf{M}k$ iff for all $t < k$, A is true at $\mathbf{M}t$.

The connectives F, P, S, U, H, G are part of the temporal predicate language and so to give models for this new enriched language we must give a system of models $\{\mathbf{M}k\}$ with the earlier-later relation $<$ on the indices (moments of time). Notice that F, P, S, U, H and G talk about time 'from within'.

When we have a sentence like A(John) ('John attends class') we cannot give it a truth value without knowing the time, i.e. the value is in a model $\mathbf{M}k$:

$$\mathbf{M}k \vDash A(\text{John}) \quad \text{iff} \quad \text{John attended class at time } k.$$

In the same way we cannot give a value to FA(John). ('John will attend class'), unless we know the time. Again we write

$$\mathbf{M}k \vDash FA(\text{John}) \quad \text{iff} \quad \text{'John will attend class' was true at time } k.$$

The connectives F, P etc. are here to allow the model k to talk about other models in other times.

The reader may ask why we use this indirect and seemingly weak way of talking about time (via P, F etc.) rather than quantify directly over time using the first method. The answer is that it is not always weak, it has a different structure and scope and can be extremely useful to many application areas, as many of the papers in this book show.

The third approach is the axiomatic approach. We write axioms and rules for the temporal language to capture all the theorems of the logic, the sentences true in any system of models $\{\mathbf{M}k\}$. The axiomatic approach is mainly connected with the second method, that of temporal connectives. However it is as relevant to the first method.

We hinted earlier that first-order predicate logic with the ability to refer to time, as in e.g. $A^*(t, x)$, only looks like first-order predicate logic but may not be exactly so. Let us elaborate. Classical model theory is related to classical provability or theorem-proving via the completeness theorem. So for example to know whether $Data \vDash B$ (in first-order semantics) we use a proof system or a theorem prover to prove B from $Data$. In our special case the quantifiers in $Data$ are restricted; t ranges over time and x ranges over students.

We cannot use an ordinary proof system or a well-known theorem prover for first-order classical logic to check whether B follows from the data. The completeness theorem tells us that $Data \nvdash B$ iff there exists a first-order model of the data in which B is false.

Here 'a model' means *any model, including* those in which $\forall t$ ranges over students and $\exists x$ ranges over times. Since we restrict the type of model allowed, we may get a new notion of $\vDash$ (semantic consequence) and we may need a new proof system or theorem prover. In fact, the new $\vDash$ may not even be arithmetical.

It can be proved that the monadic predicate temporal logic with time represented by the natural numbers is not arithmetical. The fact that we are committed to quantifying over the natural numbers as time gives us the power to define the natural model of arithmetic. I will not go into details here, but give one example to illustrate. Let

$$A = \forall x \exists t \forall s (t < s \to A^*(s, x))$$

$$B = \exists t \forall s \forall x (t < s \to A^*(s, x)).$$

A is the data. A says that for every student (every x, as x ranges over students) there is a time after which the student always attends the lecture. B is the query. B says that there is a time after which all students attend the lecture.

If we want to check whether B follows from A we cannot just take a theorem-prover for classical logic, put A in as data, B as query, push the button and let it compute. We will not get the correct answer. What we have to do is supplement the data A with axioms about the flow of time. For example we can have in predicate logic the predicates $T(u)$ for 'u is a moment of time', and $D(u)$ for 'u is an element of the domain' and we can write the following set of axioms:

(1) (a) $\forall u (D(u) \vee T(u))$
     (b) $\sim \exists u (D(u) \wedge T(u))$
(2) $\forall u, v, w \, (T(u) \wedge T(v) \wedge T(w) \wedge u < v \wedge v < w \to u < w)$
     (transitivity of $<$)
(3) $\forall u \sim [u < u \wedge T(u)]$
     (irreflexivity of $<$).
(4) $\exists u (T(u) \wedge \forall v (T(v) \to u < v))$
     (existence of a first point of time)
(5) $\forall u (T(u) \to \exists v (T(v) \wedge u < v \wedge \sim \exists w (u < w \wedge w < v)))$
     (every point has an immediate successor.)

(6) $\forall u(T(u) \wedge \exists v(T(v) \wedge v < u) \rightarrow \exists v(T(v) \wedge v < u \wedge \sim \exists w(v < w \wedge w < u)))$
(any point which is not the first point has an immediate predecessor)
(7) $\forall u, v(T(u) \wedge T(v) \rightarrow u < v \vee u = v \vee v < u)$.
(linearity).

Let $\triangle$ be the above set of axioms. It characterizes 'integer-like' time (i.e. discrete with a first element); there is no complete axiom system $\triangle'$ for exactly integer time.

Let $A\# = \forall x(D(x) \rightarrow \exists t(T(t) \wedge \forall s(T(s) \wedge t < s \rightarrow A^*(s, x))))$

$B\# = \exists t(T(t) \wedge \forall x \forall s(D(x) \wedge T(s) \wedge t < s \rightarrow A^*(s, x)))$

Then to ask whether B follows from A in the *integer-like, discrete flow of time with a first moment* one can ask instead whether

$$\triangle \cup \{A\#\} \vdash B\#$$

in first-order classical logic. One can therefore use a classical theorem-prover to check the answer.

Note that we can reduce the query $A?B$ to the query $\triangle, A\# ? B\#$ *only because* a complete axiomatization characterization $\triangle$ exists, for our flow of time. In the case of natural numbers time $\{1, 2, 3, \ldots\}$ *no such $\triangle$ exists!*

The above explains how ordinary many-sorted databases can manage to be first-order. For example consider a database with predicates like S(John, 1000) ('the salary of John is £1000'). The range of employees and possible salaries is not necessarily potentially infinite but very large and finite and hence can be characterized by some $\triangle$. In the case of natural numbers time we cannot say time is a large but finite set of moments. It is by nature infinite.

We saw above two ways of talking about the temporal properties of a sequence of models M1, M2, ..., one through time co-ordinates in the predicates $A^*(t, x)$, $R^*(t, x, y)$ and the other through additional connectives F, P, S, U, added to the language. We did not discuss how these models are to be actually represented on the computer and how we are to compute with them. They were described by us as mathematical objects. Each m-place predicate of the temporal language of the form $B(t, x1, \ldots, xm)$ is defined in the model by a subset of the power set of $N \times D^m$ where N is the set of natural numbers and D is the domain. This presentation of B is highly non-computational. If we want to do computational temporal logic we must find effective ways of presenting the temporal information on the computer. Fariba Sadri's paper in this volume falls into this category. We will now lead the discussion towards her paper. Our paper is intended to be intermediate between the mathematical and computational approach.

We begin with an example. Figure 1 describes a model of attendance and sitting, focussing on individuals, John, p and q.

	7	
John sit next to q	6	John attends
John sits next to q	5	John attends
John sits next to p	3	John attends
John sits next to p	2	John attends
	1	John attends

Fig. 1

The usual mathematical representation of this model is by means of two functions:

$$\bar{A}: \mathbb{N} \times D \rightarrow \{0, 1\}$$

$$\bar{R}: \mathbb{N} \times D \times D \rightarrow \{0, 1\}.$$

where:

$\bar{A}(m, x) = 1$ iff x attends at time m

$\bar{R}(m, x, y) = 1$ iff x sits next to y at time m.

In practice, there is a better way of representing what happens in Fig. 1, as shown in Fig. 2. The database is then just a set of entries $\{e1, \ldots, e8\}$, where we have listed the entries which are crucial to us, namely the starting of attendance and non-attendance and sitting next to. It is easy to construct the models from the above data. We need a recipe for it. Here is an example of a rule **R**1:

R1: $\bar{A}(m, J) = 1$ if there is an entry in the list of the form $A(k, J)$ where $k \leq m$ and there are no entries in the list of the form $\sim A(k', J)$ where $k \leq k' \leq m$.

$\sim R(7, J, q)$	7	$\sim A(7, J)$
	6	
$R(5, J, q)$	5	$A(5, J)$
$\sim R(4, J, p)$	4	$\sim A(4, J)$
	3	
$R(2, J, p)$	2	
	1	$A(1, J)$

Fig. 2

$\bar{A}(m, J) = 0$ if there is an entry in the list of the form $\sim A(k, J)$, where $k \leqslant m$, and there are no entries in the list of the form $A(k', J)$, where $k \leqslant k' \leqslant m$.

Let us list the entries.

$$e1 = A(1, J)$$

$$e2 = \sim A(4, J)$$

$$e3 = A(5, J)$$

$$e4 = \sim A(7, J)$$

$$e5 = R(2, J, p)$$

$$e6 = \sim R(4, J, p)$$

$$e7 = R(5, J, q)$$

$$e8 = \sim R(7, J, q).$$

We can notice that in this particular model, one cannot sit next to anyone if one is not attending. Thus we can delete entries **e6** and **e8**. Furthermore, if one is sitting next to someone then one is attending therefore we can delete entry **e3**.

What is left is

$$A(1, J)$$

$$R(2, J, p)$$

$$\sim A(4, J)$$

$$R(5, J, p)$$

$$\sim A(7, J).$$

There are two additional rules now on how to build the model from the entries, namely:

R2: $\bar{A}(m, J) = 1$ if $\exists x[\bar{R}(m, J, x) = 1]$

R3: $\bar{R}(m, x, y) = 0$ if $\bar{A}(m, x) = 0$, or if $\bar{A}(m, y) = 0$.

We must add that rules **R2** and **R3** take precedence over rule **R1**.

A *Prolog* program can be written to describe how to build the model from the entries. Furthermore, if we ask a query from the model (i.e. Is it true that Q holds in the model?) we can possibly compute the answer *directly* from the entries. If we continue developing this theme we will get a system of computational temporal logic; a way of programming temporal information in Prolog. The Event Calculus mentioned in Sadri's paper is another such example. This explains the relationship of the Event Calculus to our scheme.

There is another feature involved with the Event Calculus. Going back to our list of entries, we want the rules **R1**, **R2**, **R3**, ... involved in constructing the sequence of models from the entries to be of an intuitive nature and compatible with rules *humans use to supplement partial information*.

In other words if the entries were considered as partial information about the world, a human would complete the information also using rules **R1**, **R2** etc. To be more specific, the entry 'John attends class at time 1' was used by us as a key entry to begin the period of John attending. A human with the above partial information may reasonably assume that John continues to attend unless further information comes in. Thus rule **R1** is a good 'coding' rule.

Here is a bad rule:

'If John attends at time t, John also attends at time $t + 2$, $t + 4$, etc.'

It may be a good coding for a specific model but not necessarily what we use to supplement partial information.

Rule **R1** tells us that in case of entries of the form

$$t \qquad\qquad\qquad\qquad s$$
$$\text{attend} \qquad\qquad\qquad \text{attend}$$

there is attendance in the interval $[t, s]$. This intuitively corresponds to how we would supplement partial information.

If the above two entries come from a 'code' describing a sequence of models, then we would expect a point u of non-attendance between t and s because we always choose our 'key-coding points' by recording beginnings and endings of intervals of attendance.

The advantage of using entries is that the information about the entry can be partial. We may know that we have an entry involving attendance but not whose and when. Since we cannot write an entry of the form $A(?, ?)$ it is more convenient to describe $\mathbf{e} = {\sim} A(m, J)$ as $\mathbf{e}$ with

$$\sim = \text{begin not } (\mathbf{e})$$

$$A = \text{predicate}(\mathbf{e})$$

$$m = \text{time}(\mathbf{e})$$

$$J = \text{actor}(\mathbf{e}).$$

This way we can put $\mathbf{e}$ as an entry with $\text{predicate}(\mathbf{e}) = A$.

The above is beginning to look more and more like the Event Calculus ($\mathbf{e}$ are called events). However it is really a *Database Calculus*. Time plays no essential part here at all. We can have a database map of cities and connecting roads. To 'code' the data by entries we proceed as follows:

we choose some key entries and some intuitive non-monotonic rules which help us reconstruct the original map. These rules must be intuitive in the sense that a human would use them to construct a full map from partial information. We also need an intuitive preference system to tell us which rules are stronger than others: for this I would use Nute's (1986) *defeasible reasoning*.

The following is a language which we call **HFP** (*Hereditarily Finite Predicates*) which allows us to talk about entries and write programs about entries. This language can be used to express the Event Calculus or other temporal systems described in Sadri's paper.

The language contains the classical connectives $\sim$, $\wedge$, $\vee$, $\rightarrow$ and the variables $\{x, y, z, \ldots\}$ and the quantifiers $\forall$ and $\exists$. There is a list of atomic predicates of the form $R_n^m(\psi_1, \ldots, \psi_m, x_1, \ldots, x_n)$, where ψ_i are formula variables and x_j are term variables. There are function symbols of the form $f_n^m(\psi_1, \ldots, \psi_m, x_1, \ldots, x_n)$, where ψ_i are formula variables and x_j are term variables. We define now the notions of a *wff* and a *term* of the language **HFP**.

(1) **truth** is a formula with no free variables.

(2) x is a term with x free.

(3) If $\psi_i(x_1^i, \ldots, x_{k(i)}^i)$, $i = 1, \ldots, m$, are formulas with free variables $\{x_1^i, \ldots, x_{k(i)}^i\}$ and $F_j(\bar{x}_1, \ldots, \bar{x}_{r(j)})$, $j = 1, \ldots, n$, are terms with $\{\bar{x}_1^j, \ldots, \bar{x}_{r(j)}^j\}$ as free variables and R_n^m is a predicate symbol with (m, n) places and f_n^m is a function symbol with (m, n) places then:

   (a) $R_n^m(\psi_1, \ldots, \psi_m, F_1, \ldots, F_n)$ is a formula with the free variables $(\bar{x}_j^i, x_s^t, i = 1, \ldots, m, j = 1, \ldots, k(i), t = 1, \ldots, n, s = 1, \ldots, r(t)\}$.

   (b) $f_n^m(\psi_1, \ldots, \psi_m, F_1, \ldots, F_n)$ is a term with the same free variables as in (a).

(4) If ψ_1, ψ_2 are wffs then so are $\sim\psi_1, \psi_1 \wedge \psi_2, \psi_1 \rightarrow \psi_2, \psi_1 \vee \psi_2$. $\sim\psi_1$ has the same free variables as ψ_1, and $\psi_1 \wedge \psi_2, \psi_2 \vee \psi_2$ and $\psi_1 \rightarrow \psi_2$ have a set of free variables which is the union of the sets of free variables of ψ_1 and ψ_2.

(5) If $\psi(x, y_i)$ is a formula with free variables $\{x, y_i\}$ then $\forall x \psi$, $\exists x \psi$ are formulas with the free variables $\{y_i\}$.

To give some examples consider the **Hold** predicate. This has the form **Hold**$(\psi, t) = $ 'ψ is true at time t.'

We can now write **Hold(Hold**$(\psi, t), s))$

'At time s it was true that ψ was true at t';

compare this sentence with **Thought(Thought(ψ, 0), 5)**

'At time 5 people thought that at time 0 it was thought that ψ was true';

consider **Thought F(ψ, t)** which reads:

'At time t, ψ was considered true in the future.'

Thus we can write a rule:

If at any time you expect to get a budget for a Research Assistant, then hire one. (It may be that you never get the budget!).

$$\forall t[\textbf{Thought F}(budget, t) \rightarrow \textbf{Hire}(t)].$$

The purpose of the above discussion was to set the scene for what we are going to do in this paper, namely extend *Prolog* with temporal features. The readers interested in a discussion on the expressive power of temporal logics, on how to define them, on their axiomatizations, model theory and their general mathematical and logical properties are referred to my forthcoming books.

1 INTRODUCTION

Our problem for this paper is to start with Horn clause logic with the ability to talk about time through time co-ordinates; and see what expressive power in term of connectives (P, F, G, H, etc) is needed to do the same job. We then extend *Prolog* with the ability to compute directly with these connectives. The final step is to show that the new computation defined for P, F, G, H is really ordinary *Prolog* computation modified to avoid skolemization.

Consider now a Horn clause written in predicate logic. Its general form is of course $\bigwedge atoms \rightarrow atom$. If our atomic sentences have the form $A(x)$ or $R(x, y)$ or $Q(x, y)$ then these are the atoms one can use in constructing the Horn clause.

Let us extend our language to talk about time by following the first approach; namely, we can add time points, and allow special variables t, s to range over a flow of time $(T, <)$, (T can be e.g. the set of integers) and write instead of $Q(x, y)$, $Q^*(t, x, y)$ (also written as $Q(t, x, y)$, abusing notation), read:

'$Q(x, y)$ is true at time t.'

We allow the use of $t < s$ to mean 't is earlier than s'. Recall that we do not allow mixed atomic sentences like $x < t$ or $x < y$ or $A(t, s, x)$ because these would read like e.g.

'John loves 1980 at 1979' or
'John < 1980' or
'John < Mary.'

Assume we have organized our Horn clauses in such a manner; what kind of time expressive power do we have?

Notice that our expressive power is potentially increased. We are committed, when we write a formula of the form $A(t, x)$, that t ranges over time and x over our domain of elements. Thus our model theory for classical logic (or Horn clause logic) does not accept *any* model for $A(t, x)$, but only models in which $A(t, x)$ is interpreted in this very special way.

Meanwhile let us examine the syntactical expressive power we get when we allow for this two-sorted system and see how it compares with ordinary temporal and modal logics, with the connectives P, F, G, H.

When we introduce time variables t, s and the earlier–later relation relation into the Horn clause language we are allowing ourselves to write more atoms. These can be of the form

$$A(t, x, y)$$

$$t < s$$

(as we mentioned earlier, $A(t, s, y)$, $x < y$, $t < y$, $y < t$ are excluded).

When we put these new atomic sentences into a Horn clause we get the following possible structures for Horn clauses. $A(t, x)$, $B(s, y)$ may also be **truth**.

(a0) $A(t, x) \wedge B(s, y) \rightarrow R(u, z)$
Here $<$ is not used.
(a1) $A(t, x) \wedge B(s, y) \wedge t < s \rightarrow R(u, z)$
Here $t < s$ is used in the body but the time variable u is not the same as t, s in the body.
(a2) $A(t, x) \wedge B(s, y) \wedge t < s \rightarrow R(t, z)$
Same as (a1) except the time variable u appears in the body as $u = t$.
(a3) $A(t, x) \wedge B(s, y) \wedge t < s \rightarrow R(s, z)$
Same as (a1) with $u = s$.
(a4) $A(t, x) \wedge B(s, y) \rightarrow R(t, z)$
Same as (a0) with $u = t$, i.e. the variable in the head appears in the body.

The other two forms (b) and (c) are obtained when the head is different. (b) for time independence and (c) for a pure $<$ relation.

(b) $A(t, x) \wedge B(s, y) \rightarrow R1(z)$
(b') $A(t, x) \wedge B(s, y) \wedge t < s \rightarrow R1(z)$.
(c) $A(t, x) \wedge B(s, y) \rightarrow t < s$.
(d) $A(1970, x)$, where 1970 is a *constant* date.

Let us see how ordinary temporal logic with additional connectives can express directly, using the temporal connectives, the logical meaning of the above sentences. Note that if time is linear we can assume that one of $t < s$ or $t = s$ or $s < t$ always occurs in the body of clauses because for linear time:

$$\vdash \forall t \forall s [t < s \vee t = s \vee s < t].$$

and hence $A(t, x) \wedge B(s, y)$ is equivalent to

$$(A(t, x) \wedge B(t, y)) \vee (A(t, x) \wedge B(s, y) \wedge t < s) \vee (A(t, x) \wedge B(s, y) \wedge s < t).$$

Ordinary temporal logic over linear time allows the following connectives:

Fq, read: 'q will be true'.

Pq, read: 'q was true'

$\Diamond q = q \vee Fq \vee Pq$

$\Box q = \sim \Diamond \sim q$

$\Box q$ is read: 'q is always true'.

If $[A](t)$ denotes, in symbols, the statement that A is true at time t, then we have:

$$[Fq](t) \equiv \exists s > t([q](s))$$

$$[Pq](t) \equiv \exists s < t([q](s))$$

$$[\Diamond q](t) \equiv \exists s([q](s))$$

$$[\Box q](t) \equiv \forall s([q](s)].$$

Let us see now how to translate into temporal logic the Horn clause sentences mentioned above.

Case (a0)

Statement (a0) reads:

$$\forall t \forall s \forall u [A(t, x) \wedge B(s, y) \to R(u, z)]$$

If we push the quantifiers inside we get

$$\exists t A(t, x) \wedge \exists s B(s, y) \to \forall u R(u, z)$$

which can be written in the temporal logic as:

$$\Diamond A(x) \wedge \Diamond B(y) \to \Box R(z).$$

If we do not push the $\forall u$ quantifier inside we get $\Box(\Diamond A(x) \wedge \Diamond B(y) \to R(z))$.

Case (a1)

The statement (a1) can be similarly seen to read (we do not push $\forall t$ inside).

$$\forall t\{A(t, x) \wedge \exists s[B(s, y) \wedge t < s] \to \forall u R(u, z)\}$$

which can be translated as: $\square\{A(x) \wedge FB(y) \to \square R(z)\}$. Had we pushed $\forall t$ to the antecedent we would have got

$$\exists t[A(t, x) \wedge \exists s(B(s, y) \wedge t < s)] \to \forall u R(u, z)$$

which translates into

$$\Diamond[A(x) \wedge FB(y)] \to \square R(z).$$

Case (a2)

The statement (a2) can be rewritten as

$$\forall t[A(t, x) \wedge \exists s(B(s, y) \wedge t < s) \to R(t, z)]$$

and hence it translates to $\square(A(x) \wedge FB(y) \to R(z))$.

Case (a3)

Statement (a3) is similar to (a2). In this case we push the external $\forall t$ quantifier in and get

$$\forall s[\exists t[A(t, x) \wedge t < s] \wedge B(s, y) \to R(s, z)]$$

which translates to

$$\square[PA(x) \wedge B(y) \to R(z)].$$

Case (a4)

Statement (a4) is equivalent to:

$$\forall t[A(t, x) \wedge \exists s B(s, y) \to R(t, z)]$$

and it translates to:

$$\square(A(x) \wedge \Diamond B(y) \to R(z)).$$

Case (b)

The statement (b) is translated as

$$\Diamond(A(x) \wedge \Diamond B(y)) \to R1(z).$$

Case (*b'*)

The statement (b') translates into:

$$\Diamond(A(x) \wedge FB(y)) \rightarrow R1(z).$$

Case (*c*)

Statement (c) is a problem. It reads $\forall t \forall s[A(t, x) \wedge B(s, y) \rightarrow t < s]$. We do not have direct connectives (without negation) to express it. It says for any two moments of time t and s if $A(x)$ is true at t and $B(y)$ true at s then $t < s$. If time is linear then $t < s \vee t = s \vee s < t$ is true then we can write the conjunction

$$\sim \Diamond(A(x) \wedge PB(y)) \wedge \Diamond(A(x) \wedge B(y)).$$

Without the linearity of time how do we express the fact that t *should be* $<$-related to s? We certainly have to go beyond the connectives $P, F, \Diamond, \Box$ that we have allowed here. We will not do that; see my forthcoming book.

Case (*d*)

$A(1970, x)$ involves a constant, naming the date 1970. The temporal logic will also need a propositional constant *1970*, which is true *exactly* when the time is 1970, i.e.

$$\mathbf{M}t \vDash 1970 \quad \text{iff} \quad t = 1970.$$

Thus (d) will be translated as $\Box(1970 \rightarrow A(x))$. *1970* can be read as the proposition 'The time now is 1970' (cf. Galton in this volume, p. 190).

The above examples showed what temporal expressions we can get by using Horn clauses with time variables as an *object* language. We are not discussing here the possibility of 'simulating' Temporal Logic in Horn clause by using Horn clause as a *metalanguage*. Horn clause logic can do that to any logic as can be seen from Hodges (1985).

2 TEMPORAL PROLOG

We can now define the syntax of *Temporal Prolog*. Having added the connectives $P, F, \Box$ to the language of *Prolog* we can define recursively the following notions:

Databases
Ordinary Clauses
Always Clauses
Heads
Bodies
Goals

Definition 2.1

(1) A *Program* (or *Database*) is a set of Clauses.
(2) A *Clause* is either an Ordinary Clause or an Always Clause.
(3) An *Always Clause* is $\square A$ where A is an Ordinary Clause.
(4) An *Ordinary Clause* is a Head or an $A \rightarrow H$, where A is a Body and H is a Head.
(5) A *Head* is either an atomic formula or FA or PA where A is a conjunction of Ordinary Clauses.
(6) A *Body* is either an atomic formula, a conjunction of bodies, an FA or PA where A is a body.

We also refer to an always clause $\square A$ as having a head, namely the head of A. A *goal* is any body.

Examples

$P[F(FA(x) \land PB(y)) \land A(y)) \land A(y) \land B(x)] \rightarrow$
$$P[F(A(x) \rightarrow FP(Q(z) \rightarrow Q(y)] \text{ is an ordinary clause.}$$
So is $a \rightarrow F[(b \rightarrow Pq) \land F(a \rightarrow Fb)]$, but not $a \rightarrow \square b$.

First let us check the expressive power of this *Temporal Prolog*. Consider

$$a \rightarrow F(b \rightarrow Pq)$$

This is an acceptable clause. Its predicate logic reading is:

$$\forall t[a(t) \rightarrow \exists s > t[b(s) \rightarrow \exists u < s \ q(u)]]$$

Clearly it is more directly expressive than the Horn clause *Prolog* with time variables. Ordinary *Prolog* can rewrite the above as

$$\forall t(a(t) \rightarrow \exists s(t < s \land (b(s) \rightarrow \exists u(u < s \land q(u))))$$

which is equivalent to

$$\forall t(\exists s \exists u[a(t) \rightarrow t < s \land (b(s) \rightarrow u < s \land q(u)))))$$

If we skolemize with $\mathbf{so}(t)$ and $\mathbf{uo}(t)$ we get the clauses.

$$\forall t[(a(t) \rightarrow t < \mathbf{so}(t)) \land (a(t) \land b(\mathbf{so}(t)) \rightarrow \mathbf{uo}(t) < \mathbf{so}(t)) \land$$
$$(a(t) \land b(\mathbf{so}(t)) \rightarrow q(\mathbf{uo}(t))]$$

The following are representations of some of the problematic examples mentioned in the previous section.

(a1) $\square(A(x) \land FB(y) \rightarrow \square R(z))$

This is not an acceptable always clause but it can be equivalently written as

$$\square(\Diamond(A(x) \land FB(y)) \rightarrow R(z)).$$

(a2) $\Box(A(x) \land FB(y) \to R(z))$.

(b') $\Diamond(A(x) \land FB(y)) \to R1(z)$

(b) can be written as the conjunction below using the equation
$\Diamond q = Fq \lor Pq \lor q$:

$$(A(x) \land FB(y) \to R1(z)) \land$$

$$(F(A(x) \land FB(y)) \to R1(z)) \land$$

$$(P(A(x) \land FB(y)) \to R1(z))$$

(a2) can be similarly written.

(c) can be written as

$$\forall t, s(A(x)(t) \land B(y)(s) \to t < s)$$

This is more difficult to translate. We need negation here and write

$$\Box(A(x) \land PB(y) \to \textbf{false})$$

$$\Box(A(x) \land B(y) \to \textbf{false})$$

3 THE COMPUTATION PROCESS FOR TEMPORAL PROLOG FOR BRANCHING AND LINEAR TIME

We assume here that time is $(T, <)$ with $<$ irreflexive and transitive but not necessarily linear.

A database is composed of ordinary clauses and always clauses. We write ordinary clauses as

$$Ci = B'i \to H'i$$

and always clauses as

$$Dj = \Box(B''j \to H''j).$$

The letters $B \to H$ will schematically denote either the $B' \to H'$ of an ordinary clause or the $B'' \to H''$ of an always clause $\Box(B'' \to H'')$. Let **P** be a database and Q a goal. Assume different clauses in **P** use different variables. The goal uses different variables as well. We define inductively the notion 'Q succeeds from **P** with some substitution θ', denoted by '$\textbf{P}?Q = 1(\theta)$'; note that θ is the successful substitution! The following describes how we conceive the computation.

Given a database **P**, all clauses in **P** are understood as universally quantified. Given a goal G, all variables in G are read as existential variables to be instantiated. Thus we are looking for a substitution θ for the variables of

G such that, in the appropriate modal or temporal logic:

(universal closure of all clauses in **P**) $\vdash G\theta$.

The computation is conceived theoretically as if the correct guess for θ has been made at the beginning and all we have to do is to show that the computation is successful by giving the correct sequence of universal instantiations to clauses of **P** so that $G\theta$ succeeds.

This point of view is essential because of problems with Skolem functions in temporal and modal logics; $\theta 1$ below must substitute constants for all variables.

(A) *Case of Q atomic*
(1) $\mathbf{P}?Q = 1(\theta)$ if for some head $H \in \mathbf{P}$ and some $\theta 1$, $H\theta 1 = Q\theta$
(2) $\mathbf{P}?Q = 1(\theta)$ if for some ordinary clause $Bi \to Hi \in \mathbf{P}$ and some $\theta 1$, $Hi\theta 1 = Q\theta$ and $\mathbf{P}?Bi = 1(\theta 1)$
(3) $\mathbf{P}?Q = 1(\theta)$ if for some clause $\Box(Bj \to Hj) \in \mathbf{P}$, $Hj\theta 1 = Q\theta$ and $\mathbf{P}?Bj = 1(\theta 1)$

(B) *Case of Q a conjunction $A \wedge B$*

$\mathbf{P}?A \wedge B = 1(\theta)$ iff $\mathbf{P}?A = 1(\theta)$ and $\mathbf{P}?B = 1(\theta)$.

(C) *Case of $Q = FA$*
(C1) $\mathbf{P}?FA = 1(\theta)$
if for some $B \to FH \in \mathbf{P}$ and $\theta 1$ the following two subcomputations both succeed.

Sub Computation 1
Either
(1a) {all always clauses of **P**} $\cup$ {$PC\theta 1 \mid C$ an ordinary clause of **P**} $\cup$ {$H\theta 1$}$?A = 1(\theta)$
or
(1b) {Same data}$?FA = 1(\theta)$.

Sub Computation 2
$\mathbf{P}?B = 1(\theta 1)$

Explanation for the computation (C1):
The database (Program) **P** is conceived as what is true now.

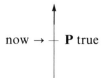

$$\text{now} \to \quad \mathbf{P} \text{ true}$$

We find a clause $B \to FH \in \mathbf{P}$, whose universal closure is true now. We are asking: Is $A\theta$ true in the future?

i.e. we are asking:

Is $FA\theta$ true now?

If $H\theta1 \vdash A\theta$, then if $B\theta1$ is true now then $FH\theta1$ is true now and hence $FA\theta$ is true now.

So we have to check first whether $H?A = 1(\theta)$ can succeed, and then check whether $\mathbf{P}?B$ succeeds.

When we ask $H?A$, we can add to H also PC for any clause C and all the always clauses. Because time is transitive we know that if we succeed with FFA it follows that FA succeeds. Hence we also ask $H?FA$. This explains possibility (1b).

(C2) $\mathbf{P}?FA = 1(\theta)$ if

either C2(*) or C2(**) succeed, where:

        C2(*) succeeds if for some $\square(B \to FH) \in \mathbf{P}$, subcomputations (1) and (2) below all succeed for some $\theta1$.

        C2(**) succeeds if for some $\square(B \to H) \in \mathbf{P}$ subcomputation (1) succeeds and (2b) succeeds for some $\theta1$:

        Sub Computation 1: (For both cases (*) and (***)):

        (1a) Either $\{H\theta1\} \cup \{$all always clauses of $\mathbf{P}\} \cup \{PC|C$ a clause of $\mathbf{P}\}$ $?A = 1(\theta)$

        or

        (1b) $\{$same data$\}$ $?FA = 1(\theta)$

        Subcomputation 2 Either (2a) or (2b) for case (*) and (2b) only for case (**).

        (2a) $\mathbf{P}?B = 1(\theta1)$

        (2b) $\mathbf{P}?FB = 1(\theta1)$

        (i.e. $B \vee FB$ succeeds from $\mathbf{P}$ for case (*)).

    Note that for case C2(**) subcomputation (2a) is not required for the success of the goal of (C2).

(D): *Case of $Q = PA$*: This is the mirror image of the case of $Q = FA$ described above.

Remarks:

We can read $\mathbf{P}?A \vee B = 1(\theta)$ as a disjunction $P?A = 1(\theta)$ or $P?B = 1(\theta)$. See Definition 4.2 on negation by failure for a mathematical formulation for the computation process in terms of trees.

Example 3.1

Data (1) $a \to Fb$

       (2) $\square(b \to Fc)$

       (3) a

Query: Fc

To succeed we ask, using datum (2), *c* from *c* and *Fb* from the data using computation C2(**). To continue, we ask, using (1), *b* from *b* and *a* from the data, which succeeds. The computation succeeds.

Example 3.2

Data
(1) $\Box(A \rightarrow FB)$
(2) *FA*.
Query: *FB*

The computation succeeds. We ask $B \vee FB$ from $\{A \rightarrow FB\}$ and then *FA* from $\{FA\}$. This is computation C2(**).

Figures 3 and 4 show two computation trees for the examples above: We have included the words '*purpose*: success' in the boxes to anticipate the use of negation by failure in which case the purpose could be to fail.

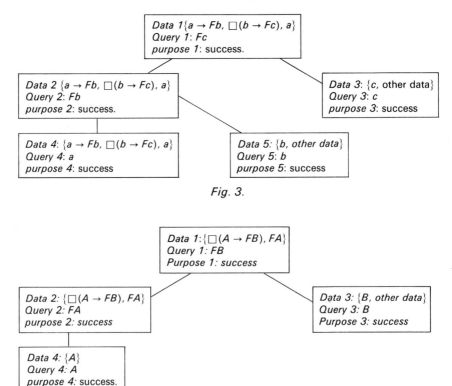

Fig. 3.

Fig. 4.

So far in this section we have described the computation for branching transitive time. We have to add more rules for linear time. We will do that in a continuation paper. We now proceed to show the soundness of the computation rules relative to a system of temporal logic.

Definition 3.3

The temporal logic for transitive branching time has the following axioms and rules, in the language with P and F:
First define:

$$Hq = \sim P \sim q; \quad Gq = \sim F \sim q$$

$$\Diamond q = q \vee Fq \vee Pq, \quad \Box q = \sim \Diamond \sim q.$$

(1) All theorems and rules of predicate logic.

(2) $\dfrac{\vdash A}{\vdash HA}$; $\dfrac{\vdash A}{\vdash GA}$; $\dfrac{\vdash A, \vdash A \to B}{\vdash B}$

(3) $G(A \to B) \to (GA \to GB)$
$H(A \to B) \to (HA \to HB)$
$GA \to GGA$
$HA \to HHA$

(4) $A \to GPA$
$A \to HFA$

The following additional axioms yield linear time

(5) $FA \wedge FB \to F(A \wedge FB) \vee F(B \wedge FA) \vee F(A \wedge B)$
$PA \wedge PB + P(A \wedge PB) \vee P(B \wedge PA) \vee P(A \wedge B)$

Remark
The following are theorems of this logic:

$$\forall x \Box A(x) \to \Box \forall x A(x)$$

$$\forall x GA(x) \to G\forall x A(x)$$

$$\forall x HA(x) \to H\forall x A(x)$$

Theorem 3.4
Let **P** be a database of *Temporal Prolog* and let Q be a goal; then $\mathbf{P}?Q = 1(\theta)$ implies:

$(\mathbf{UP}) \vdash Q\theta$, where **UP** denotes the universal closure of $\bigwedge \mathbf{P}$.

Proof: By induction on the computation steps of Q from **P**.
(A) (1) If for some $H \in \mathbf{P}$ and some $\theta 1$, $H\theta 1 = Q\theta$, then clearly $(\mathbf{UP}) \vdash Q\theta$.

(2) If for some ordinary clause $Bi \rightarrow Hi \in \mathbf{P}$, and some $\theta 1$, $Hi\theta 1 = Q\theta$
and $\mathbf{P}?Bi = 1(\theta 1)$ then by the induction hypothesis $(\mathbf{UP}) \vdash Bi\theta 1$
but
$(\mathbf{UP}) \vdash Bi\theta 1 \rightarrow Hi\theta 1$
hence $(\mathbf{UP}) \vdash Hi\theta 1$
i.e. $(\mathbf{UP}) \vdash Q\theta$.

(3) If for some always clause $\square(Bj \rightarrow Hj) \in \mathbf{P}$, and some $\theta 1$, $Hj\theta 1 = Q\theta$
and $\mathbf{P}?Bj = 1(\theta 1)$ then by the induction hypothesis $(\mathbf{UP}) \vdash Bj\theta 1$.
but
$(\mathbf{UP}) \vdash \forall x \square(Bj \rightarrow Hj)$
hence
$(\mathbf{UP}) \vdash \square\forall x(Bj \rightarrow Hj)$
$(\mathbf{UP}) \vdash \forall x(Bj \rightarrow Hj)$
$(\mathbf{UP}) \vdash Bj\theta 1 \rightarrow Hj\theta 1$
hence
$(\mathbf{UP}) \vdash Hj\theta 1 = Q\theta$.

(B)   Clearly $\mathbf{P}?A \wedge B = 1$ iff $\mathbf{P}? A = 1$ and $\mathbf{P}?B = 1$,
iff $\mathbf{P} \vdash A$ and $\mathbf{P} \vdash B$, iff $\mathbf{P} \vdash A \wedge \mathrm{B}$.

(C1) Assume $\mathbf{P}?FA = 1(\theta)$ and that the computation succeeds because for
some $B \rightarrow FH$ in $\mathbf{P}$ and some $\theta 1$ we have
(a) $\{H\theta 1\} \cup \{$always clauses of $\mathbf{P}\} \cup \{PC\theta 1 | C$ an ordinary clause of $\mathbf{P}\}$
$?A \vee FA = 1(\theta)$
and
(b) $\mathbf{P}?B = 1(\theta 1)$.
Then by the induction hypothesis we get

(1) $H\theta 1 \wedge \bigwedge ($always clauses of $\mathbf{P}) \wedge \bigwedge_{C \in \mathbf{P}} PC\theta 1 \vdash (A \vee FA)\theta$
and

(2) $(\mathbf{UP}) \vdash B\theta 1$
We want to show $(\mathbf{UP}) \vdash FA\theta$
From (1) and (2) we get that:
$\bigwedge \{$always clauses of $\mathbf{P}\} \vdash H\theta 1 \wedge PC\theta 1 \rightarrow (A \vee FA)\theta$ and
$(\mathbf{UP}) \vdash B\theta 1$
Also, $(\mathbf{UP}) \vdash B\theta 1 \rightarrow FH\theta 1$ and $(\mathbf{UP}) \vdash C\theta 1$, for each ordinary clause
of $\mathbf{P}$.
Since $\vdash C \rightarrow GPC\theta 1$ we get that
$(\mathbf{UP}) \vdash G(H\theta 1 \rightarrow (A \vee FA)\theta)$
Since $\mathbf{P} \vdash FH\theta 1$ we get $\mathbf{P} \vdash (FA \vee FFA)\ \theta$ and therefore $(\mathbf{UP}) \vdash$
$FA\theta$.

(C2) $\mathbf{P}?FA = 1(\theta)$ if either
(*) for some $\square(B \rightarrow H) \in \mathbf{P}$
or (**) for some $\square(B \rightarrow FH) \in \mathbf{P}$,
the following hold, for some $\theta 1$.

(1) $\{H\theta 1\} \cup \{$all always clauses of $\mathbf{P}\} \cup \{PC | C \in \mathbf{P}\} ? (A \vee FA) = 1(\theta)$
(2) $\mathbf{P}?B\theta 1 \vee FB\theta 1 = 1$
    Hence by the induction hypothesis
    (1) $(\mathbf{UP}) \vdash H\theta 1 \wedge \bigwedge \{$always clauses$\} \wedge \bigwedge PC\theta 1 \vdash (A \vee FA)\theta$
    (2) $(\mathbf{UP}) \vdash B\theta 1 \vee FB\theta 1$
    From (1) we get that $\mathbf{P} \vdash G(H\theta 1 \rightarrow (A \vee FA)\theta)$
    from (2), since we have either (*) or (**) where
    (*): $\mathbf{P} \vdash \Box(B \rightarrow H)$
    (**): $\mathbf{P} \vdash \Box(B \rightarrow FH))$
    we get that $\mathbf{P} \vdash FH\theta 1$ and hence
    $\mathbf{P} \vdash F(A \vee FA)\theta,$
    and therefore
    $\mathbf{P} \vdash FA\theta.$
(D) The case of $\mathbf{P}?PA$ is the mirror image of the case of $\mathbf{P}?FA$.

Further Remarks
In the description of the temporal computation in this section we required that the substitution $\theta 1$ substitutes constants for all variables. This restriction is motivated by the following example.

Example 3.5

Data $\mathbf{P}$: $A(a)$
        $A(x) \rightarrow FB(y)$
Query Q: $F(B(a) \wedge B(b))$.

If we let $\theta 1$ substitute $x = a$, but leave y as it is, then we get the two further subcomputations:

(1) $\{A(a), A(x) \rightarrow FB(y)\} ? A(a)$ and
(2) $\{B(y)\} ? B(a) \wedge B(b)$.

Viewed on its own, the second computation will succeed because we can substitute $y = a$ and $y = b$ separately. However, it should not succeed because we should not be allowed to substitute different values for y. The data clause given to us in the database $\mathbf{P}$ is $\{\forall x \, \forall y (A(x) \rightarrow FB(y))\}$, and reasoning forward from $A(a)$ we get $\forall y \, FB(y)$. The future point in which $B(y)$ is true depends on y, and may be different for different y's. Thus although we can conclude $FB(a) \wedge FB(b)$, we cannot conclude $F(B(a) \wedge B(b))$. The restriction on $\theta 1$ forces $\theta 1$ to substitute $y = $ constant, and therefore we cannot carry out the above erroneous computation.

    In practice we do not have to guess right at the beginning of the computation what the successful $\theta 1$ is. We can substitute for y a dummy constant $\bar{y}$, deciding later what actual constant it is supposed to be. Thus the second computation becomes:

$$\{B(\bar{y})\} \, ? \, B(a) \, \wedge \, B(b).$$

If we choose $\bar{y} = a$, in order to succeed with the left conjunct $?B(a)$, we cannot then continue and succeed with $?B(b)$, by letting $\bar{y} = b$, since $\bar{y}$ is a constant and has already been chosen to be a.

In a practical *Prolog* environment one can implement $\{B(\bar{y})\}$ as a *list*, with $\bar{y}$ a variable. In this situation, once $\bar{y}$ is unified with the constant a, $\bar{y}$ becomes a everywhere in the computation where it occurs.

To be more detailed, a Database is composed in *Prolog* of some clauses and a list. A metapredicate describes the unification and computation process by saying that we first try to unify with heads of clauses in the data and then with heads of clauses in the list. The *Prolog* machine will keep the clause variables universal but the list variables will become instantiated everywhere they occur. The subcomputations will create new databases and new lists according to the computation rules. We thus create a metapredicate interpreted in *Prolog* of the form *Succeed (Data, list, goal)*.

To describe this computation theoretically we can assume that we have two types of variables; $\{x, y, z, \ldots\}$ for universal variables and $\{\bar{x}, \bar{y}, \bar{z}, \ldots\}$ for constants to be chosen. A database $\mathbf{P}$ can contain both types of variables in its clauses and a query Q contains *only* the variables with a bar ($\bar{x}, \bar{y}$ etc). The meaning of a successful computation of the typical form

$$\mathbf{P}(x, \bar{y}) \, ? \, Q(\bar{z})$$

is

$$\vdash \exists y, z (\forall x \mathbf{P} \rightarrow Q)$$

In other words, we ask whether we can substitute constants for $\bar{y}, \bar{z}$ such that $\forall x \mathbf{P} \vdash Q$. According to this terminology, all substitutions $\theta 1$ and θ are to constants and terms containing names and bar variables $\{\bar{x}, \bar{y},\}$ and all substitutions to $\bar{x}, \bar{y}$ must be persistent and be carried out simultaneously everywhere in all the databases in which they occur, which are involved in the computation. This type of variable was first introduced in Gabbay and Reyle (1984). I am indebted to Fariñas del Cerro for pointing out that the problem arises here as well.

In fact, with the device of two types of variables, nothing stops us from writing clauses of the form

(1) $A(x) \rightarrow F(\forall y B(y) \wedge C(z))$.
(2) $A(a)$.

To ask Fq we first can ask $?A(a)$ from the database and continue with the second computation;

$$\{B(y), C(\bar{z})\}\,?\,q.$$

Here we put $B(y)$ in the database rather than $B(\bar{y})$, because $\forall y B(y)$ was under F; $\bar{z}$ is a constant to be chosen.

We would also like to explain why we have not allowed clauses of the form $a \to b$, within the head of the clause. There is no theoretical reason for it. The reason is computational. We will treat the case of $\to$ and **false** in heads of clauses in a continuation paper.

Example 3.6

Imagine a database with

$$a1 \to \Box A$$

$$a2 \to \Box(A \to C)$$

$$a3 \to \Box D3$$

$$a4 \to \Box D4$$

$$a1$$

$$a2.$$

If our query is $\Box C$ the correct computation rule is to try *each finite* set of clauses $ai \to \Box \text{head}(i)$ and ask C from $\{\text{head}(i)\}$ and $\bigwedge ai$ from the database.

Thus if we choose too few clauses (e.g. $\{a1 \to \Box A\}$, the subcomputation $\{A\}\,?\,C$ will not succeed. If we take too many, e.g. $\{a1 \to \Box A, a2 \to \Box(A \to C),$ $a3 \to \Box D3\}$, then although $\{A, A \to C, D\}\,?\,C$ does succeed, the other subcomputation $?\,a1 \wedge a2 \wedge a3$ will not succeed from the database.

As an extension of *Prolog*, it is not practical. However, if we are looking for a theorem prover for full modal and temporal logics, then this can be done. We can extend our method to a full theorem prover by adding computation rules for $\to$ and falsity ($\mathbf{P}\,?\,A \to B$ is reduced to $\mathbf{P} \cup \{A\}?B$) in the spirit of Gabbay and Reyle (1984). We can overcome the problems with Skolem functions and possible worlds of different domains by using ideas of Section 5 below. This we will do in a continuation paper.

4   NEGATION AND NEGATION AS FAILURE

We can add $\neg A$ to mean negation as failure. It is up to us to understand the meaning of $\neg A$. We explain the possibilities through an example. Let $\mathbf{P1}$ be:

$$a \to F(\neg b \to FC)$$

$$a$$

$$b$$

Query: ask for $?FFC$

We go to two subcomputations. We ask for **P1**?*a*, which succeeds and ask for

$$\mathbf{P2}?\,FC, \quad \text{where} \quad \mathbf{P2} = \{\neg b \to FC\}$$

To do the latter computation we again go to two subcomputations. We ask for *C* from $\{C\}$ = **P3** and then ask for $\neg b$ from the then current database **P2**, i.e. we ask **P2**? $\neg b$, which succeeds, since $b \notin$ **P2**. If we read $\neg b$ as failure from **P1** then we fail. Out intuition is to ask $\neg b$ relative to the current world, because that is what we would do if $\neg b$ were a positive b^*.

The full computation with negation by failure can be described formally as follows: we describe it for the propositional case.

Definition 4.1

Consider a language with propositional atoms, $\wedge$, $\to$, $\vee$, $\neg$, F, P, $\square$ and $\diamondsuit$. We define the notions of databases Ordinary Clauses, Always Clauses, Heads, Bodies and Goals.

(1) A *Program* (or *Database*) is a set of Clauses.
(2) A *Clause* is either an Ordinary Clause or an Always Clause.
(3) An *Always Clause* is $\square A$ where *A* is an Ordinary Clause.
(4) An *Ordinary Clause* is a Head or an $A \to H$, where *A* is a Body and *H* is a Head.
(5) A *Head* is either an atomic formula or *FA* or *PA* where *A* is a conjunction of Ordinary Clauses.
(6) A *Body* is either an atomic formula, a conjunction of bodies, an *FA* or *PA* or $\neg A$ where *A* is a body.

We also refer to an always clause $\square A$ as having a head, namely the head of *A*. A goal is any body.

Definition 4.2

Let **P** be a database and *G* a goal. We define the notion of a *labelled computation tree* for the success or the finite failure of *G* from **P**. If *G* succeeds we write **P**?*G* = 1 and if *G* finitely fails we write **P**?*G* = 0. Let $(T, \leqslant, 0, V)$ be a labelled computation tree, such that $0 \leqslant t$, for all $t \in T$. *V* is a labelling function giving each $t \in T$ a triple $(\mathbf{P}(t), G(t), x(t))$, where $\mathbf{P}(t)$ is a database, $G(t)$ is a goal, and $x(t) = 0$ or $x(t) = 1$. When $x(t) = 1$ we want the goal to succeed. When $x(t) = 0$ we want it to fail. $(T, \leqslant, 0, V)$ is a labelled computation tree for **P**?*G* = *x* iff the following conditions are satisfied.

(0) $V(0) = (\mathbf{P}, G, x)$.
(1) If $t \in T$ is an endpoint and $x(t) = 1$ then $G(t)$ is an atom *q* and $q \in \mathbf{P}(t)$, or $\square q \in \mathbf{P}(t)$.

(2) If $t \in T$ is an endpoint and $x(t) = 0$ then either (a) or (b), where:
  (a) $G(t)$ is an atom q and q is neither the head of an ordinary clause, nor is there in $\mathbf{P}(t)$ any always clause with head q (i.e. of the form $\Box(body \to q)$);
  (b) $G(t)$ has the form FA (or PA) and there are no clauses with heads of the form FD (respectively PD).

(3) If t is not an endpoint and $x(t) = 1$ and $G(t)$ is an atom q, then t has exactly one immediate successor s in the tree with $\mathbf{P}(s) = \mathbf{P}(t)$, $x(s) = 1$ and either
$$G(s) \to q \in \mathbf{P}(t) \quad \text{or} \quad \Box(G(s) \to q) \in \mathbf{P}(t).$$

(4) If t is not an endpoint and $x(t) = 0$ and $G(t)$ is an atom q, then t has $s1, \ldots, sk$ as immediate successors in the tree with $\mathbf{P}(si) = \mathbf{P}(t)$, $x(si) = 0$ and for some $0 \leqslant m \leqslant k$, the clauses $G(si) \to q$ for $i \leqslant m$, and the clauses $\Box(G(si) \to q)$ for $m \leqslant i \leqslant k$, are exactly all the clauses of the above form in $\mathbf{P}(t)$.

(5) If t is not an endpoint and $x(t) = 1$ and $G(t) = A1 \wedge A2$, then t has exactly two immediate successors $s1$ and $s2$ with $\mathbf{P}(si) = \mathbf{P}(t)$, $x(si) = 1$ and $G(si) = Ai$.

(6) If t is not an endpoint, $x(t) = 0$ and $G(t) = A1 \wedge A2$, then t has exactly one immediate successor s and for some $i \in \{1, 2\}$, $\mathbf{P}(s) = \mathbf{P}(t)$, $x(si) = 0$ and $G(si) = Ai$.

(7) If t is not an endpoint and $G(t) = \neg A$, then t has exactly one immediate successor and $G(s) = A$, $x(s) = 1 - x(t)$ and $\mathbf{P}(s) = \mathbf{P}(t)$.

(8) If t is not an endpoint and $G(t) = FA$ and $x(t) = 1$, then t has exactly two immediate successors $s1$ and $s2$ and one of (a) ... (e) holds, where
  (a) $\mathbf{P}(s1) = \mathbf{P}(t)$, $x(s1) = 1$ and for some H
    $G(s1) \to FH \in \mathbf{P}(t)$ and
    $x(s2) = 1$ and $\mathbf{P}(s2) = \{$all always clauses of $\mathbf{P}(t)\} \cup \{PC|C$ and an ordinary clause of $\mathbf{P}(t)\} \cup \{H\}$
    and $G(s2) = A$
    and $x(s2) = 1$
  (b) Same as (a) above except that $G(s2) = FA$.
  (c) Same as (a) above except that the condition $G(s1) \to FH \in \mathbf{P}(t)$ is replaced by $\Box(G(s1) \to FH) \in \mathbf{P}(t)$.
  (d) Same as (c) above except that $G(s2) = FA$.
  (e) Same as (b) above except that the condition $G(s1) \to FH \in \mathbf{P}(t)$ is replaced by $\Box(G(s1) \to H) \in \mathbf{P}(t)$.

(9) If t is not an endpoint and $G(t) = PA$ and $x(t) = 1$ then t has exactly two immediate successors $s1$ and $s2$ and one of the cases (a'), ..., (e') holds, where (a') ... (e') are the mirror images of (a), ..., (e) of (8) above (where P and F are interchanged).

(10) If t is not an endpoint and $G(t) = FA$ and $x(t) = 0$, then t has $s1, \ldots, sk$
as its immediate successors, and for some $1 \leqslant m \leqslant n \leqslant k$ and some wffs
$D1, \ldots, Dk$, $Di(i \leqslant m)$ are exactly all clauses of the form $\square(Bi \to Hi)$ in
$\mathbf{P}(t)$, $D(m + 1), \ldots, D(n)$ are exactly all clauses of the form $\square(Bi \to FHi)$
in $\mathbf{P}(t)$, $D(n + 1), \ldots, D(k)$ are exactly all clauses of the form
$Bi \to FHi \in \mathbf{P}(t)$, and for each $1 \leqslant i \leqslant k$ either (*) or (**) holds:

(*) $G(si) = Bi$, $\mathbf{P}(si) = \mathbf{P}(t)$ and $x(si) = 0$.
(**) $G(si) = A \vee FA$, for $1 \leqslant i \leqslant k$ (except when $m < i < n$, in which
case $G(si) = FA$), and $x(si) = 0$ and $\mathbf{P}(si) = \{$all always clauses of
$\mathbf{P}(t)\} \cup \{Hi\} \cup \{PC | C$ an ordinary clause of $\mathbf{P}(t)\}$.

(11) If t is not an endpoint and $G(t) = PA$ and $x(t) = 0$, then a condition
similar to that of (10) holds with P and F interchanged.

(12) If t is not an endpoint and $G(t) = A1 \vee A2$ and $x(t) = 0$, then t has
exactly two immediate successors $s1$ and $s2$ with $\mathbf{P}(si) = \mathbf{P}(t)$ and
$G(si) = Ai$ and $x(si) = 0$.

(13) If t is not an endpoint and $G(t) = A1 \vee A2$ and $x(t) = 1$, then t has
exactly one immediate successor s and $\mathbf{P}(s) = \mathbf{P}(t)$ and $G(s) = \{A1, A2\}$
and $x(s) = 1$.

Definition 4.3

Define:

(a) $\mathbf{P}?G$ *succeeds* if $\mathbf{P}?G = 1$ has a finite computation tree.
(b) $\mathbf{P}?G$ *finitely fails* if $\mathbf{P}?G = 0$ has a finite computation tree.

Let us explain the intuition behind the definition of the labelled computation tree. Our (success) computation rules all have the form

$$\mathbf{Success}(Data, Goal) \text{ if } \bigwedge_i \mathbf{Success}(Data\ i, Goal\ i)$$

where $Data\ i$, $Goal\ i$ are obtained from the original $Data$ and $Goal$. The above
is a *meta-Horn clause* describing part of the *Temporal Prolog* computation
given in Section 3. For example, if the database $\mathbf{P}1$ contains the clause
$a \to Fb$ and the goal is Fc then one of our rules says that Fc succeeds from $\mathbf{P}1$
if $\mathbf{P}2 = \{$always clauses of the data$\} \cup \{PC | C \in \mathbf{P}1\} \cup \{b\} ? c$ succeeds and
$\mathbf{P}1?a$ succeeds. Thus we can write the meta-Horn clause corresponding to this
computation rule as:

$\mathbf{Success}$ ($\mathbf{P}1$, Fc) if for some $a \to Fb \in \mathbf{P}1$, $\mathbf{Success}$ ($\mathbf{P}2$, c) and $\mathbf{Success}$ ($\mathbf{P}1$, a).

The Horn clause meta-rules for the meta-predicate $\mathbf{Success}$ ($\mathbf{P}$, G) are exactly
the computation rules of our *Temporal Prolog* expressed as Horn clauses for
that meta-predicate.

The labelled computation tree is the ordinary computation tree for the *Prolog* computation of the predicate **Success**, where we write (**P**, *G*, 1) instead of **Success** (**P**, *G*). As examples, see the computation diagrams given in Section 3 for examples 3.1 and 3.2.

The failure predicate **Failure(P**, *G*) records exactly when the predicate **Success(P**, *G*) finitely fails in the meta-Prolog sense. The definition of labelled computation tree for failure is nothing more than a complete and explicit record of the *Prolog* failure of the meta-predicate **Success**, for the particular meta-program which is our temporal *Prolog* computation defined in Section 3.

The above explains what definition 4.3 does. It explicitly defines finite failure for the predicate **Success**.

Incidentally, I cannot resist putting myself in the shoes of my esteemed colleague R. A. Kowalski and say that if all our *Temporal Prolog* computation is just the meta-predicate **Success(P**, *G*) why bother extending *Prolog* formally with *P*, *F*, $\diamond$, $\square$ and defining a computation for them? Why not remain in ordinary *Prolog* and, whenever any temporal features are needed, use the Horn clause program for the predicate **Success**? This is a good point.

The answer is subtle and can never be conclusive. It is a matter of presentation; similarly, why not simulate Pascal features in Basic? One of the later sections of this paper will give some hints. There is a difference between the use of a simulated meta-language and the use of a direct object language. We are *extending Prolog* as an object language. The predicate **Success** in *Prolog* is a *simulation* of our *Temporal Prolog*. Indeed that is how we are likely to implement *Temporal Prolog*.

5   COMPUTATION WITHOUT SKOLEMIZATION

The purpose of this section is to show that our *Temporal Prolog* computation can be regarded as classical logic computation without skolemization.

This idea is best explained by an example.

Consider the database **P1**:

(1)   **Rich**(x) $\wedge$ $\forall z$ **Sleep**(x, z) $\rightarrow$ $\exists u$ **Hate**(u, x)
(2)   **Hate**(u, x) $\rightarrow$ $\exists y$ (**Friend**(y, u) $\wedge$ **Kill**(y, x)).
(3)   **Rich**(John).
(4) $\forall x$ **Sleep**(John, x)

The data are self-explanatory ('sleep' means 'sleep with'). Written properly as a pure Horn clause program we get:

(1) $\forall x \exists z \exists u$ [**Rich**(x) $\wedge$ **Sleep**(x, z) $\rightarrow$ **Hate**(u, x)]
(2) $\forall u \forall x \exists y$ [**Hate**(u, x) $\rightarrow$ **Friend**(y, u) $\wedge$ **Kill**(y, x)]

(3)   **Rich**(John).

(4) $\forall x$ **Sleep**(John, x)

After skolemizing we get the *Prolog* program:

(1)   **Rich**(x) $\wedge$ **Sleep**($x, f1(x)$) $\rightarrow$ **Hate**($f2(x), x$)

(2a) **Hate**(u, x) $\rightarrow$ **Friend**($g(u, x), u$)

(2b) **Hate**(u, x) $\rightarrow$ **Kill**($g(u, x), x$)

(3)   **Rich**(John)

(4)   **Sleep**(John, x).

Consider the goal **Kill**(v, John). This goal succeeds.

Suppose we want to avoid skolemizing. We do not want to use $f1, f2$ and $g(u, x)$ which may complicate things for us. Let us rewrite the database in an equivalent form.

(1) $\forall x[$**Rich** $(x) \rightarrow \exists z[$**Sleep**($x, z$) $\rightarrow \exists u$ **Hate**(u, x)]]

(2) $\forall x \forall u$ [**Hate**(u, x) $\rightarrow \exists y$ [**Friend**(y, u) $\wedge$ **Kill**(y, x)]]

(3)   **Rich**(John)

(4) $\forall x$ **Sleep**(John, x).

Our goal is $\exists v$ **Kill**(v, John).

Consider the following computation:

(1) I want to show the existence of v such that **Kill**(v, John). What gives me existence in the database?

(2) Clause 2 gives me existence of a y. Fix y, would my goal follow from what exists i.e. from
     the database $\cup \{$**Friend**(y, u) $\wedge$ **Kill**(y, x)$\}$?
     The answer is yes.
     Let $v = y$ (the y which exists) and $x =$ John.

(3) What would give me this existence (of y)? I need to get **Hate**(u, John), and any u will do. What can give existence? From clause (1) we get existence of z. Fix z and see whether $\exists u$ **Hate**(u, John) can succeed, i.e. ask $\exists u$ **Hate**(u, John) from the database together with the clause (*)$\{$**Sleep**(x, z) $\rightarrow \exists u$ **Hate**(u, x)$\}$.
     This succeeds if $x =$ John, and **Sleep**(John, z) succeeds. However, to get (*) itself we must also ask the antecedent of clause (1) from the database, namely **Rich** (John).

Figure 5 shows the computation: **P** is the database, $\exists v$ **Hate**(v, John) is the goal.

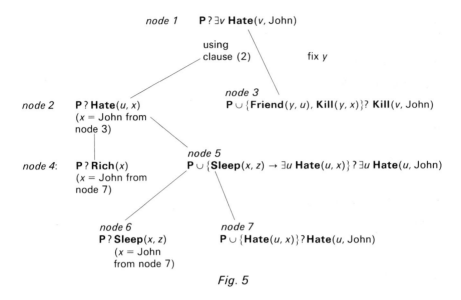

Fig. 5

Let us compare this computation with the temporal logic example 3.2.

Data
(1) $\Box(A \rightarrow FB)$

(2) FA

Query
FB.

Let us write it in predicate logic.

Data
(1) $\forall t[A(t) \rightarrow \exists y(t < y \land B(y))]$

(2) $\exists z\,(\text{now} < z \land A(z))$

Query $\exists u\,(\text{now} < u \land B(u))$.

Implicit in the computation is the transitivity of $<$, namely

(3) $s < t \land t < r \rightarrow s < r$.

Computation:
(1) What can give

$$\exists u\,(\text{now} < u \land B(u))?$$

Clause (1) gives existence, ask therefore for

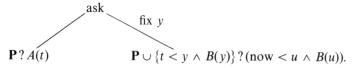

$\mathbf{P}?A(t)$ $\mathbf{P} \cup \{t < y \wedge B(y)\}?(\text{now} < u \wedge B(u)).$

Use clause (3), make $s = \text{now}$, $v = u$ and $u = y$. The right-hand node succeeds, we must ask the left-hand node $A(t)$ which also succeeds.

   The above was a complex computation, especially because variables were fixed or identified across nodes in the computation tree. For the moment at least let us identify a fragment of predicate logic for which we want to do this computation.

Definition 5.1

We define the notion of E-database, E-clause, E-body and E-head (E for Existential).

(1) An *E-database* is a set of E-clauses.
(2) An *E-clause* is either an atomic formula or an E-head H or a $B \to H$, where B is an E-body and H is an E-head.
(3) An *E-head* is of the form $\exists y\, A(y)$, where $A(y)$ is a conjunction of E-clauses or an E-head.
(4) An *E-body* is a conjunction of atomic formulas.
(5) A *goal* is any E-body.

The variables in the goal are read existentially.

Definition 5.2   E-computation

Let $\mathbf{P}$ be an *E*-database and let $\mathbf{G}$ be a goal; we define the notion 'the goal G succeeds from $\mathbf{P}$ under the substitution θ'. We use the notation

$$\mathbf{P}?G = 1(\theta)$$

for this.

(1) $\mathbf{P}?A \wedge B = 1(\theta)$ iff
    $\mathbf{P}?A = 1(\theta)$ and $\mathbf{P}?B = 1(\theta)$.
(2) $\mathbf{P}?A = 1(\theta)$ if for some $B \to \exists y\, H(y)$ and some $\theta 1$ we have $\mathbf{P}?B\theta 1 = 1(\theta)$
    and $\mathbf{P} \cup \{H(y0)\theta 1\}? A = 1(\theta)$ where $y0$ is a *new* Skolem constant.
(3) $\mathbf{P}?Q = 1(\theta)$ for Q atomic if for some atomic $H \in \mathbf{P}$ and some $\theta 1$,
    $H\theta 1 = Q\theta$.

It is obvious that the computation is sound. Rule (2) is really forward modus ponens. If we translate $a \to Fb$ as $a(x) \to \exists y(x < y \wedge b(y))$ then *Temporal Prolog* computation from $a \to Fb$ is the same as E-computation from $a(x) \to \exists y(x < y \wedge b(y))$.

We hope that the above discussion has shown the connection of Temporal Logic Programming with computing without skolemizing; what remains for us to check is the expressive power of E-clauses. We will see that in the case of monadic predicate logic (i.e. in which all atomic formulas are unary) the expressive power is quite strong. This is our next task.

Let us proceed with some examples:

Example 5.3

Consider the formula $\forall x \exists y[A(x) \to B(y)]$ where A and B are atomic. To prepare this formula for computation we have to skolemize. We will get:

$$\forall x[A(x) \to B(f(x))]$$

with f the Skolem function.

We can also push $\exists y$ inside the original formula and get

$$\forall x[A(x) \to \exists y B(y)]$$

Now push $\forall x$ inside and get

$$\exists x A(x) \to \exists y B(y);$$

now pull $\exists y$ and then $\exists x$ outside and get

$$\exists y \forall x[A(x) \to B(y)].$$

If we skolemize now we get:

$$\forall x[A(x) \to B(c)],$$

where c is a constant.

In the above example the exchange of the quantifier order worked because A and B were unary predicates and so the occurrences of x and y were neatly separated. We do not have something like $R(x, y)$ in the monadic case.

Let us look at a more general example.

Example 5.4

Consider the formula:

$$\forall x \exists y[[A1(x) \to B1(y)] \wedge [A2(x) \to B2(y)]]$$

In this example it is more difficult to push $\exists y$ inside because $\exists$ does not go across a conjunction. However, because we are dealing with monadic predicates we have that $\psi = \exists y[[A1(x) \to B1(y)] \land [A2(x) \to B2(y)]]$ is logically equivalent to ψ^*, where:

$$\psi^* = [A1(x) \to \exists y\, B1(y)] \land [A2(x) \to \exists y\, B2(y)] \land [A1(x) \land A2(x)$$
$$\to \exists y[B1(y) \land B2(y)]]$$

More generally, if $\psi(x)$ is $\exists y \bigwedge\limits_{i=1}^{m} [Ai(x) \to Bi\,(x,y)]$ then it is equivalent to $\psi^*(x)$ where

$$\psi^*(x) = \bigwedge_{\substack{J \subseteq \{1,\ldots,m\} \\ J \neq \varnothing}} \left[\bigwedge_{j \in J} Aj(x) \to \exists y\left[\bigwedge_{j \in J} Bj(x,y)\right]\right]$$

To prove the above we show that in any model $\forall x\, [\psi(x) \Leftrightarrow \psi^*(x)]$ is true.

Proof
(a) Assume for any x that $\psi(x)$ is true (in a model). Show that $\psi^*(x)$ is true. ψ^* is a conjunction. We must show that each conjunct is true. Let $J \subseteq \{1,\ldots,m\}$; we must show that

$$\bigwedge_{j \in J} Aj(x) \to \exists y\left[\bigwedge_{j \in J} Bj(x,y)\right]$$

is true. Hence,

If $\bigwedge\limits_{j \in J} Aj(x)$ is not true then the implication *is* true. If $Aj(x)$ is true for each $j \in J$, we use the fact that $\psi(x)$ is true, and choose a y promised by $\psi(x)$. Thus y satisfies $\bigwedge\limits_{i=1}^{m} [Ai(x) \to Bi(x,y)]$; hence, since $Aj(x)$ are all true for $j \in J$, we get that $Bj(x,y)$ are all true for $j \in J$, and so $\exists y\left[\bigwedge\limits_{j \in J} Bj(x,y)\right]$ is true.

We have thus shown that if $\psi(x)$ is true in any model so is $\psi^*(x)$.

(b) Assume $\psi^*(x)$ is true in a model. We show that $\psi(x)$ is true. We define a Skolem function $f(x)$ such that, in the model,

$$\forall x \bigwedge_{i=1}^{m} [Ai(x) \to Bi\,(x,f(x))]$$

is true. Hence,

$$\exists y \bigwedge_{i=1}^{m} [Ai(x) \to Bi(x,y)]$$

will be true for the choice of $y = f(x)$.

Take any x. Let $J \subseteq \{1,\ldots,m\}$ be a maximal set of indices for which $\bigwedge\limits_{j \in J} Aj(x)$ is true. Thus $Ai(x)$ is true iff $i \in J$. Since ψ^* holds in the model,

$\bigwedge_{j \in J} Aj(x) \to \exists y \bigwedge_{j \in J} Bj(x, y)$ holds, and since $\bigwedge_{j \in J} Aj(x)$ is true, there exists a y

such that $\bigwedge_{j \in J} Bj(x, y)$ is true. Let $f(x)$ be such a y. Thus f is defined for every

x. If all $Ai(x)$ are false let $f(x)$ be some fixed value $y0$. Then

$\psi = \exists y \bigwedge_{i=1}^{m} [Ai(x) \to Bi(x, y)]$ is true in the model.

To show this take $y = f(x)$; clearly, by construction of f, if $Ai(x)$ is true then $Bi(x, f(x))$ is true. This concludes the proof and also concludes example 5.4.

Example 5.5

Notice that in example 5.4 we have shown that if $\psi(x) = \exists y \bigwedge_{i=1}^{m} [Ai(x) \to Bi(x, y)]$ then $\psi(x)$ is logically equivalent to $\psi^*(x)$, where

$$\psi^*(x) = \bigwedge_{J \subseteq \{1, \dots, m\}} \left[\bigwedge_{j \in J} Aj(x) \to \exists y \bigwedge_{j \in J} Bj(x, y) \right]$$

Therefore $\forall x\, \psi^*(x)$ is the same logically as $\forall x\, \psi(x)$. Let us check what happens to $\forall x\, \psi^*(x)$ in case $Bi(x, y)$ is independent of x, i.e. is of the form $Bi(y)$.

Then we get the form

$$\forall x\, \psi^*(x) = \forall x \bigwedge_{\substack{J \subseteq \{1, \dots, m\} \\ J \neq \varnothing}} \left[\bigwedge_{j \in J} Aj(x) \to \exists y \left[\bigwedge_{j \in J} Bj(y) \right] \right].$$

Pushing $\forall$ across the $\wedge$ in ψ^* we get:

$$\forall x\, \psi^*(x) = \bigwedge_{J} \forall x \left[\bigwedge_{j \in J} Aj(x) \to \exists y \bigwedge_{j \in J} Bj(y) \right];$$

pushing $\forall$ into the antecedent of each implication we get:

$$\forall x\, \psi^*(x) = \bigwedge_{J} \left[\exists x \bigwedge_{j \in J} Aj(x) \to \exists y \bigwedge_{j \in J} Bj(y) \right].$$

We can now pull out $\exists y$ and $\exists x$ and get

$$\forall x\, \psi^*(x) = \bigwedge_{J} \exists y\, \forall x \left[\bigwedge_{j \in J} Aj(x) \to \bigwedge_{j \in J} Bj(y) \right].$$

We thus get our final equivalence of (*) and (**)

$$(*): \forall x\, \exists y \bigwedge_{i=1}^{m} [Ai(x) \to Bj(y)]$$

is equivalent to

$$(**): \bigwedge_{\substack{J \subseteq \{1, \ldots, m\} \\ J \neq \varnothing}} \exists y \, \forall x \left[\bigwedge_{j \in J} Ai(x) \rightarrow \bigwedge_{j \in J} Bj(y) \right].$$

It is thus possible in the monadic predicate logic to exchange the order of quantifiers $\forall \exists$ and $\exists \forall$.

Notice that if the conjuncts in (*) are Horn clauses and (*) is a datum in a Horn clause database i.e. Ai are conjunctions of atoms (not containing y) and $Bi(y)$ are atoms, then (**) is also a conjunction of Horn clauses, because (**) is equivalent to (***), where

$$(***): \bigwedge_{\substack{J \subseteq \{1, \ldots, m\} \\ J \neq \varnothing}} \exists y \, \forall x \bigwedge_{j \in J} \left[\bigwedge_{j \in J} Aj(x) \rightarrow Bj(y) \right]$$

Thus if (*) is in a Horn clause database, we can replace it by (***). So instead of skolemizing (*) in the Horn clause database with a Skolem function $y = f(x)$, we can skolemize (***) with constants, but no function symbols.

Let us now go back to our monadic predicate logic. Assume the language contains monadic predicates $A(x)$ and also propositions q without variables. Let ψ be an arbitrary sentence of this logic. Write ψ in a conjunctive prenex normal form. Thus ψ is written as

$$(Q1x1, \ldots, Qkxk) \bigwedge_{i=1}^{m} \left(\bigvee_{j=1}^{n} Aij(xij) \vee \bigvee_{j=1}^{n} bij \right)$$

where Aij is either an atom or the negation of an atom with a variable xij, and bij is just an atomic proposition or the negation of an atomic proposition. The quantified variables are of course from among the free variables of the body of ψ, i.e. $\{x1, \ldots, xk\} \subseteq \{xij\}$. We have allowed here atomic formulas without variables, to help with our induction.

Assume the inner quantifier $(Qkxk)$ is $\exists y$. Let $\psi(x1, \ldots, x(k-1))$ be

$$\exists y \bigwedge_{i=1}^{m} \left[\bigvee_{j=1}^{n} Aij \vee \bigvee_{j=1}^{n} bij \right].$$

We can push $\exists y$ across the conjunction as follows. Write the disjunction $\bigvee_j Aij \vee \bigvee_j bij$ as $Ai(x1, \ldots) \rightarrow Bi(y)$ where $Bi(y)$ contains all disjuncts with y free and all the bij, and $Ai(x1, \ldots)$ is the negation of the disjunction of the other disjuncts. If there are no disjuncts with y in them and no bij then one can always rewrite a disjunction $\bigvee Aij$ as $\bigwedge \neg Aij \rightarrow false$.

Thus the above form is the most general. It is always possible to rewrite to this form because we are dealing with monadic logic. *Each* variable appears in clearly defined formulas without 'sharing' with other variables, i.e. we do not have the case of $R(x, y)$ where we do not know where to put $R(x, y)$.

We can therefore write ψ as $\psi 1$

$$\psi 1(x1, \ldots) \equiv \bigwedge_i [Ci(x1, \ldots) \rightarrow \exists y\, Di(y)]$$

$Di(y)$ contains no free variables other than y. This is possible in view of the proof in example 5.4.

If the quantifier $Qkxk$ is $\forall y$, we can easily push $\forall$ across the conjunction. Then we have conjuncts of the form $\forall y\, [Ai(x) \rightarrow Bi(y)]$ and again we get

$$[Aj(xi) \rightarrow \forall y\, Bi(y)];$$

again $\forall y\, Bi(y)$ is without free variables.

We can thus prove by induction that any formula $\psi(x1, \ldots, xn)$ with xi free can be written in the following normal form:

$$\bigwedge_i Ai(x1, \ldots, xn) \rightarrow Qy\, Bi(y)$$

where $Bi(y)$ is in normal form in y and Q is $\forall$ or $\exists$.

Definition 5.6

A formula ψ is said to be in an *M-form* if

(a) ψ is in the form $Qy\, A(y)$ where $A(y)$ is a boolean combination of atomic formulas with free variable y only.
(b) ψ has the form $Qy\, B(y)$ where $B(y)$ is a boolean combination of formulas which either are atomic with y free or are in M-form.

Example 5.7

$$\forall x\, \exists y\, R(x, y)$$

can never be written in normal form because $R(x, y)$ has two free variables. For computation with run-time skolemization, see Gabbay and Reyle (1987).

6   DISCUSSION AND CONCLUSION

Our aim has been to see how Logic Programming can handle temporal phenomena in a natural way. This certainly involves translating or simulating Temporal Logics in *Prolog*, whenever possible, and extending *Prolog* with additional features. We thus have to clarify the various notions of *translation* involved. This is the purpose of this section.

Prolog can simulate any recursive or recursively enumerable set. This was

pointed out by Kowalski and is indeed easy to see. Thus any axiomatizable or recursively presented temporal logic can be 'simulated' in *Prolog*. Hodges (1985) has shown that the semantics of classical logic can be simulated in Horn clause logic by looking at Boolean-valued models. This is more like a translation than a symbolic simulation. Simulation is not the same as a direct translation. To explain the difference between a direct translation and a simulation, we give some examples. We cannot give you formal mathematical definitions at this stage. These concepts will be discussed in my (forthcoming) book on Programming in Pure Logic.

Example 6.1

Let **L1** be the extension of classical logic with a modality operator $\Box$. Let us read $\Box A$ to mean 'A is necessarily true', and endow $\Box$ with axioms to make it a transitive modality. Thus the axioms for **L1** are (besides all classical logic theorems and rules):

(a) $\dfrac{\vdash A}{\vdash \Box A}$

(b) $\vdash \Box(A \wedge B) \Leftrightarrow \Box A \wedge \Box B$

(c) $\vdash \Box A \rightarrow \Box\Box A$

(d) $\vdash \forall x \Box P(x) \rightarrow \Box \forall x P(x)$

L1 does *not* prove the reflexivity axiom

(e) $\Box A \rightarrow A$

i.e.

$$\mathbf{L1} \nvdash \Box A \rightarrow A.$$

Example 6.2

We now define a temporal logic **L2** with a 'progressive' operator IA reading:

'A is true now and around now.'

To make the semantics of IA more precise, imagine a flow of time like the rationals or real numbers:

$$-x-t-y \rightarrow$$

IA is true at t if and only if there is an interval around t in which A is true, i.e.

$$\exists x, y(x < t < y \wedge \forall u(x < u < y \rightarrow A \text{ true at } u))$$

This corresponds to some reading of the English progressive-tense e.g. 'He is walking now'.

It can be shown that **L2** entails exactly all the axioms of **L1** together with the reflexivity axiom. In other words the modal axioms (a) ... (e) axiomatize all the theorems of **L2**.

We begin by pointing to three different ways of translating **L2** into **L1**. We mean here a mapping **T**, translating each wff of **L2** into a wff of **L1**:

$$\mathbf{T}: A \mapsto \mathbf{T}(A),$$

so that for some set of wffs **S** of **L1** we have for all A

$$\mathbf{L3} \vdash A \quad \text{iff} \quad \mathbf{S} \vdash_{\mathbf{L1}} \mathbf{T}(A).$$

We define three different translations **T**, denoted by **T1, T2, T3**.

The translations are defined by induction on the structure of the formulae of **L2**, as follows:

(a) $\mathbf{T}(atom)$ = as described by the particular translation
(b) $\mathbf{T}(A \wedge B) = \mathbf{T}(A) \wedge \mathbf{T}(B)$
(c) $\mathbf{T}(\sim A) = \sim \mathbf{T}(A)$
(d1) $\mathbf{T}(A \vee B) = \mathbf{T}(A) \vee T(B)$
(d2) $\mathbf{T}(A \rightarrow B) = \mathbf{T}(A) \rightarrow \mathbf{T}(B)$
(e) $\mathbf{T}(\exists x A(x)) = (\exists x \in D) \, \mathbf{T}((A(x))$
(f) $\mathbf{T}(\forall x A(x)) = (\forall x \in D) \, \mathbf{T}(A(x))$
     where $D(x)$ is a formula of **L1**
(g) $\mathbf{T}(IA) = M(\mathbf{T}(A))$
     where $M(Q)$ is a formula of **L1** with the predicate Q.

The above is not the most general way of translating. Firstly $\wedge$, $\vee$, $\sim$, $\rightarrow$ were translated to themselves; this is not always so, see for example translation **T2** below. Secondly each A was translated into a single $\mathbf{T}(A)$. In the general case, $\mathbf{T}(A)$ can be a set of wff of **L1** and we have

$$\mathbf{L2} \vdash A \quad \text{iff} \quad \forall B \in \mathbf{T}(A)(\mathbf{S} \vdash_{\mathbf{L1}} B).$$

The following are the three different ways of translating **L2** into **L1**.

(1) *Direct Translation*
   $\mathbf{T1}(Q) = Q$, for Q atomic.
   $\mathbf{T1}(IA) = A \wedge \Box A.$

It can be shown that

$$\mathbf{L2} \vdash A \quad \text{iff} \quad \mathbf{L1} \vdash \mathbf{T1}(A).$$

(2) *Simulation*
Since **L1** extends predicate logic, a provability predicate *Demo* (P, B) can be

defined in **L1** for the provability process in **L2** of B from the data P. Suitable naming of the form $T2(A) = name\ (A) = n(A)$ can be given and we can have, for a suitable set of axioms **S** about the translation that

$$\textbf{L2} \vdash A \quad \text{iff} \quad \textbf{S} \vdash_{\textbf{L1}} Demo(\varnothing, n(A))$$

(3) *Semantic Translation*

In **L1** a special variable $t \in T$, and a special set T for time can be allocated, as well as time relation $<$. Any atom $Q(xi)$ of **L2** can be translated as $Q^*(t, xi)$ of **L1** reading '$Q(xi)$ is true at time t', and T3 can be defined as

$$T3(Q) = Q^*(t),\ Q \text{ atomic, and}$$

$$T3(IA) = \exists x\ \exists y(x < t < y \wedge \forall u[x < u < y \rightarrow T3(A)(u)])$$

The translation T3 actually defines the semantical interpretation of **L2** inside **L1**. We again have that there exists a theory **S** of **L1** such that for any A of **L2**

$$\textbf{L2} \vdash A \quad \text{iff} \quad \textbf{S} \vdash_{\textbf{L1}} T3(A).$$

S is needed to say that we are dealing with time in **L1**, that $D \cup T$ is the entire universe, and that some other organizational properties are satisfied.

The reader may think that any translation T1, T2, T3 is no better than the others, but this is not the case. Consider the following points:

Point 1

In translation T1 the addition of the axiom $\Box A \rightarrow A$ in L1 'collapses' the meaning of $T(IA)$ into $\Box A$. In translation T2 and T3 there is no effect on $T(IA)$. In other words, in a direct translation the image is not a 'module' and there are reflections from the target logic to the image.

Point 2

In translation T2 (the provability simulation), for any extension of **L2** the source can be *automatically* translated. We just modify the provability predicate. This is not true for T1 and T3. We may in the case of T1 or T3 find new translations defined from scratch, but even this cannot always be done.

Point 3

T1 and T2 when applied to themselves (i.e. **L1** = **L2**) yield practically the identity, not so for T3.

We see that properties of translations can be studied in terms of:

(1) Extendibility
(2) Reflection principles

(3) Compositions
(4) Other properties

We may be able to give a semi-categorical definition of what a direct translation is and how it is different from a simulation.

Acknowledgements

I am grateful to F. del Cerro, A. P. Galton and R. A. Kowalski for valuable criticism. This research was partially supported by Esprit Cost 13 joint project: *Logical Techniques in Knowledge Representation and Natural Language Analysis*, at Imperial College and the FNS Institute at University of Tubingen.

References

Gabbay, D. (1986), '*Executable Temporal Logic for Interactive Systems*'. Technical Report, Imperial College.

Gabbay, D. (in preparation), *Expressive Power of Temporal Logics*.

Gabbay, D. (in press), *Programming in Pure Logic: an Honest Approach*. Ellis Horwood, Chichester.

Gabbay, D. and Reyle, U. (1984), 'N-Prolog—An extension of *Prolog* with hypothetical implication', *Journal of Logic Programming* 1, 319–355.

Gabbay, D. and Reyle, U. (1987), '*Computation with Run-time Skolemization*'. Technical Report, Imperial College.

Hodges, W. (1985), '*Reducing First order Logic to Horn Logic*.' Report, Queen Mary College.

Kowalski, R. A. and Sergot, M. J. (1986), 'A logic-based calculus of events', *New Generation Computing* 4, 67–95, Springer Verlag.

Nute, D. (1986), '*LDR, A logic for defeasible reasoning*', ACMC Research Report 01-0013. University of Georgia.

INDEX